SpringerBriefs in Applied Sciences and Technology

Manufacturing and Surface Engineering

Series Editor

Joao Paulo Davim ⓘ, Department of Mechanical Engineering, University of Aveiro, Aveiro, Portugal

This series fosters information exchange and discussion on all aspects of manufacturing and surface engineering for modern industry. This series focuses on manufacturing with emphasis in machining and forming technologies, including traditional machining (turning, milling, drilling, etc.), non-traditional machining (EDM, USM, LAM, etc.), abrasive machining, hard part machining, high speed machining, high efficiency machining, micromachining, internet-based machining, metal casting, joining, powder metallurgy, extrusion, forging, rolling, drawing, sheet metal forming, microforming, hydroforming, thermoforming, incremental forming, plastics/composites processing, ceramic processing, hybrid processes (thermal, plasma, chemical and electrical energy assisted methods), etc. The manufacturability of all materials will be considered, including metals, polymers, ceramics, composites, biomaterials, nanomaterials, etc. The series covers the full range of surface engineering aspects such as surface metrology, surface integrity, contact mechanics, friction and wear, lubrication and lubricants, coatings an surface treatments, multiscale tribology including biomedical systems and manufacturing processes. Moreover, the series covers the computational methods and optimization techniques applied in manufacturing and surface engineering. Contributions to this book series are welcome on all subjects of manufacturing and surface engineering. Especially welcome are books that pioneer new research directions, raise new questions and new possibilities, or examine old problems from a new angle. To submit a proposal or request further information, please contact Dr. Mayra Castro, Publishing Editor Applied Sciences, via mayra.castro@springer.com or Professor J. Paulo Davim, Book Series Editor, via pdavim@ua.pt.

More information about this subseries at http://www.springer.com/series/10623

Manjunath Patel G. C. · Ganesh R. Chate ·
Mahesh B. Parappagoudar · Kapil Gupta

Machining of Hard Materials

A Comprehensive Approach
to Experimentation, Modeling
and Optimization

 Springer

Manjunath Patel G. C.
Department of Mechanical Engineering
PES Institute of Technology
and Management
Shivamogga, India

Mahesh B. Parappagoudar
Department of Mechanical Engineering
Padre Conceicao College of Engineering
Verna, Goa, India

Ganesh R. Chate
Department of Mechanical Engineering
KLS Gogte Institute of Technology
Belgaum, India

Kapil Gupta
Department of Mechanical
and Industrial Engineering Technology
University of Johannesburg
Doornfontein, Johannesburg, South Africa

ISSN 2191-530X   ISSN 2191-5318   (electronic)
SpringerBriefs in Applied Sciences and Technology
ISSN 2365-8223   ISSN 2365-8231   (electronic)
Manufacturing and Surface Engineering
ISBN 978-3-030-40101-6   ISBN 978-3-030-40102-3   (eBook)
https://doi.org/10.1007/978-3-030-40102-3

This Springer imprint is published by the registered company Springer Nature Switzerland AG
The registered company address is: Gewerbestrasse 11, 6330 Cham, Switzerland

Preface

Machining is one of the important product development operations. Productivity, quality, and sustainability are the three major categories of machinability indicators to determine the success of any machining operation. When it comes on machining of difficult-to-machine and/or hard materials, then selection and optimization of the appropriate machining techniques are of prime importance. Research, development, and innovations in the field lead to analyse, observe, and understand the mechanism of machining for various materials. This book provides insights into machining of hard material. It includes outcomes of the research conducted by the authors in the form of experimental results, their analysis, intelligent modelling, and optimization. The main objective of this book is to present the possibilities to engineer machining operations to facilitate the machining of various difficult-to-machine and/or hard materials. It starts with Chap. 1 where hard materials and machining processes are introduced. Chapter 2 presents review of some previous work on machining of hard materials along with discussing machining variables and techniques for experimentation, modelling, and optimization. Results of experimental investigation on machining of hard material EN31 steel, conducted by the authors, are reported, analysed, and discussed in Chap. 3. It also discusses the effect of important machining parameters on material removal rate of the process and surface roughness, cylindricity, and circularity of the machined part of EN31 steel. Chapter 4 is focussed on neural network-based intelligent modelling of the process to establish the relationship between machining parameters and responses/machinability indicators. Soft computing optimization of machining of EN31 steel is reported in Chap. 5.

The information presented and investigation results reported in this book are from the cutting-edge research conducted by the authors in this area. The authors hope that the research reported on the experimentation, modelling, and optimization would facilitate and motivate the researchers, engineers, and specialists working in the field of advanced manufacturing and materials engineering.

We sincerely acknowledge Springer for this opportunity and their professional support.

Shivamogga, India   Manjunath Patel G. C.
Belgaum, India   Ganesh R. Chate
Goa, India   Mahesh B. Parappagoudar
Johannesburg, South Africa   Kapil Gupta

Contents

Chapter 1
Introduction to Hard Materials and Machining Methods

1.1 Introduction

Machining is most widely used to transform the material into the product of desired shape and size by the mechanism of removing excess material. Machining involves group of processes, wherein the excess material is removed from the work specimen in sequential steps with the help of cutting tools (either single point or multi-point). It is to be noted that machining with a single-point cutting tool uses well-defined tool geometry (i.e. cutting edges (honed, sharp, chamfered) possessing different faces (rake, flank, etc.), whereas grinding process uses abrasive wheel with multi-point micro-cutting edges having undefined geometry [1–3]. Machining processes are widely used to finish parts of both metallic (ferrous, non-ferrous, and their alloys) and non-metallic materials (polymers, composites, wood, glass, etc.) to the desired complex geometries possessing high structural integrity and surface finish [4]. Hence, machining process is widely used in industries for manufacturing moulds and dies of casting; injection moulding, extrusion, forging, and so on; automotive, hydraulic, and other power transmission and aircraft parts [4–9].

Machining operations account approximately 15% of the total manufacturing cost, and the survey report shows that the annual expenditure associated with machining and machine tools will be around several billion euros in industrially developed countries [10, 11]. It is worthy to note that the trend shows an increase of 20–30% machining applications of aeronautical parts [12]. Rapid development in machining technology has enabled the machining industries to manufacture the products with high accuracy and precision. These technologies are broadly categorized as traditional and non-traditional machining processes. Each process has its own distinct advantages and limitations [13, 14]. Slow material removal rate, part size (i.e. thickness limitations), multi-pass machining even for thin sections, high tooling and capital investment cost, low productivity, restriction to a wide variety of materials are few shortcomings of non-traditional machining technology [14]. Indeed, for manufacturing or design engineers, the major concern is with regard to functional

© The Author(s), under exclusive license to Springer Nature Switzerland AG 2020
M. Patel G. C. et al., *Machining of Hard Materials*,
Manufacturing and Surface Engineering,
https://doi.org/10.1007/978-3-030-40102-3_1

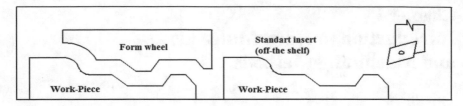

Fig. 1.1   Hard turning versus grinding process

(i.e. material processing ability, productivity, yearly production, size and shape of part geometry, surface integrity, lead time, accuracy, power and energy consumption, utilization of existing machine/machining practice, cost-effectiveness, etc.) requirement, which lie in selecting the appropriate machining process [15–18]. Hence, the machining processes are to be selected in such a way that they are technically and economically feasible and reduce the cost without altering the quality of machined part. For example, Fig. 1.1 shows the part geometry machined with grinding and turning process. The advantages of turning over grinding include no or reduced requirement of coolants, high material removal rate which will result in less set-up changes, reduced power, tool and machining cost, ability to cut thin part section, improved productivity [19, 20]. In addition, to finish part geometry the multiple turning operations can be carried out in a single set-up, whereas multiple set-ups are required in grinding [21]. An excellent surface quality can be achieved in grinding as compared to turning process. This occurs due to uncontrollable tool wear in turning process, which deteriorates the surface finish and integrity [22]. Above discussion shows a lot of ambiguity, due to complexity involved in selecting the machining process. Hence, prior background knowledge regarding the production capabilities of machining process is an important prerequisite. In addition, modern industries are selecting the machining process based on the following [19]: (a) applications, i.e. part geometry with regard to desired dimensional tolerance and surface integrity required, (b) machining lead time and productivity, (c) reduced energy consumption, (d) machining cost, (e) tool wear, and so on.

Keeping in view of machining hard materials or difficult-to-cut materials, turning process is an alternative to grinding which will yield the following significant benefits [23, 24]: (a) around 60% reduction in machining time and energy consumption, (b) versatility and flexibility to produce complex geometries with minimum errors, (c) little or complete elimination of coolant or cutting fluid, (d) environment-friendly machining practice (i.e. no problem with disposal of cutting fluids and their associated diseases). During machining of the parts using turning, approximately 97% of the mechanical energy is transformed to thermal (i.e. heat) energy. It is to be noted that around 80% of heat is in primary shear zone, out of which chips and turned parts will carry 75% and 5%, of heat, respectively. The remaining 18 and 2% of the total thermal energy will be generated at tool–chip interfaces and tool–workpiece interfaces, respectively [25]. The heat generated at interfaces due to friction phenomenon will increase the tool wear and produce deteriorated surface on machined parts. Cutting fluids are used to limit the shortcomings and associated negative impact of heat and friction between tool and workpiece. The positive impact of using cutting fluids is cooling, lubricating, and carrying the chips away from the cutting zone, which will help to improve the tool life and surface finish [26]. The said positive benefits must compensate with the negative impact of cutting fluids such as cost, working environment, and disposal of cutting fluid. It is estimated that the cost incurred with cutting fluid varies in the range of 7–17% of total manufacturing cost [27]. This cost may even raise to 30% to combat the losses of cutting fluids during machining [28]. Losses of cutting fluid are attributed to the vapourization losses, loss with machine components, leakage, maintenance for cleaning and drying system, and so on [28]. Note that the operators of machining industries face health problems, wherein majority (say 80%) of these diseases are due to negative effect of cutting fluid [29]. Although there exists a significant advantage reported on the use of cutting fluids, it is justified that the coolant does not have major significant impact on hard turning process [24]. Dry machining technology will resolve the problems (waste disposal, health issues, environmental problems, and huge cost) of cutting fluids with the utilization of advanced cutting tool materials [30]. Diniz and Micaroni [31] reported that while dry machining (without coolant) with appropriate choice of cutting condition had resulted in a better surface finish compared to wet turning process [31]. Figure 1.2 shows the qualitative comparison of hard turning and grinding process capabilities. From the detailed literature review, turning process substitutes the grinding process to finish the part geometry that is extremely difficult-to-cut materials with significant advantages in terms of flexibility, versatility, ecology, economy, and quality.

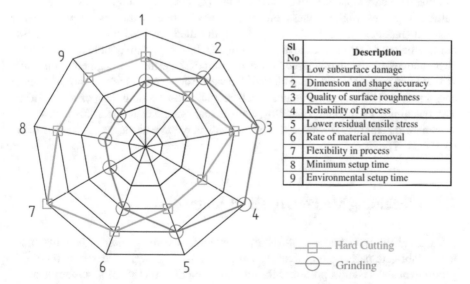

Sl No	Description
1	Low subsurface damage
2	Dimension and shape accuracy
3	Quality of surface roughness
4	Reliability of process
5	Lower residual tensile stress
6	Rate of material removal
7	Flexibility in process
8	Minimum setup time
9	Environmental setup time

□— Hard Cutting
○— Grinding

Fig. 1.2 Qualitative comparison of hard turning and grinding process capabilities

1.2   Hard Materials

Hard turning refers to the machining of hard materials or difficult-to-cut materials (usually above 45 HRC) by utilizing a single-point cutting tool. In the recent past, significant attention was paid towards the development of a wide range of hard materials for industrial applications. Industrial relevance of hard material includes abrasives and cutting tools with excellent wear and scratch resistance along with surface and chemical stability [32]. The success of conventional machining process depends on hardness of tool material, which must be much higher than the hardness of workpiece material [33]. Challenge lies in machining of such harder materials with economic and other technological benefits.

Hard materials are those which offer greater resistance to undergo deformation, indentation, or penetration. The materials (i.e. soft or hard) are generally classified based on the hardness, which can be measured and expressed as hardness number. The common methods used to measure the dimensionless parameter of hardness of materials are Brinell hardness, Rockwell hardness, Vickers hardness, Knoop hardness, and so on. Rockwell hardness test has significant benefits such as no additional optical device is required to record the hardness value, uses differential depths to limit the errors that occur due to surface imperfections, and economically and comparatively causes less damage to material part. The Rockwell hardness testing uses standard set of scales designated by a letter (i.e. A, B, C, D, E, F, … and V). The different letters will correspond to different scales and be used for recording hardness values of different materials. Higher the number recorded in the scale, higher will be the hardness values. Hard materials are those whose hardness values measured on Rockwell C scale must be above 45 HRC [34]. Materials with high hardness number value will possess high hardness. Further, materials are classified as extremely hard material (hardness number above 55 HRC) and medium hard materials (hardness number below 55 HRC and above 45 HRC) [35]. It is important to note that the hardness changes as a result of altered flow behaviour of materials, i.e. segmentation [36]. A few hard materials include white/chilled cast iron, steels (such as tool, bearing, case-hardened, hard chrome-coated, heat-treated, and so on), and super-alloys [37]. The machining of Inconel, Hastelloy, Stellite and other exotic materials is referred to as hard machining. However, these materials may have hardness values less than 45 HRC [37].

1.3   Machining Methods of Hard Materials

In the early 1980s, technological advancement took place in the field of hard machining to fabricate parts with difficult-to-cut materials. Although researchers reported grinding process is most preferred for fine finishing of parts [38, 39], numerous benefits corresponding to machining of hard materials by utilizing cutting tool should not be neglected [19, 20, 40]. Each machining process will produce its own surface texture, and the removal of layers of material from the workpiece depends on its mechanism [41]. Thereby, significant attention is required to study the various methods of machining hard materials with the cutting tools and the same are discussed below.

1.3.1 Hard Turning

Hard turning is a viable process intended to limit or replace the conventional grinding process for machining hard work materials, whose hardness value on Rockwell C scale is greater than 45. Note that the hard turning process is capable to generate the surface finish of 0.4–0.8 μm Ra, roundness of 2–5 μm, and dimensional tolerance of ±3–7 μm [34]. Hard turning with coated ceramic tool successfully generated the surface roughness of value 0.352 μm Ra [42]. Interestingly, high precision machining with CBN tool generated IT3 good dimensional tolerance with surface roughness values close to 0.1 μm Ra [43]. Studies show that CBN tool can produce better surface finish than ceramic inserts, but latter is preferred due to very low cost associated with it [44]. Hence, clear vision in selection of cutting tool material for machining the hard materials will play a vital role. In addition, appropriate choice of cutting conditions (cutting speed, feed, and depth of cut), workpiece material variables (hardness, toughness, and other mechanical properties), and tool variables (tool geometry, material and its hardness, tool overhang and its vibration, and so on) are also of significant importance in hard turning [44]. It is important to note that the parts required extremely tight tolerance, hard turning serves the purpose of pre-finishing operation as intermediate process and latter can be accomplished with finish grinding process. This cost-effective processing route will enable the industries to machine parts of automotive, bearings, die, moulds, etc. Although many researchers stated hard turning is a potential substitute for grinding process [45–48], full substitution might not be easy and possible. This might be due to the inherent process capability differences between the grinding and turning operations. The potential differences are observed from the literatures mentioned below [22, 49–53]:

1. Hard turning operations are comparatively faster, wherein the parts can be machined in a single pass and set-up even at dry cutting conditions.
2. Hard turning operations performed in a lathe will enable a greater flexibility to finish complex geometries.
3. Both rough and finish cuts can be obtained with single clamping in a lathe.
4. Material removal rate in hard turning provides economic benefits.
5. Although tooling costs are comparatively higher in turning due to rapid tool wear and failure, this negative impact may be balanced with the new developments in tool materials.
6. Multi-face and profile cuts will offer greater flexibility in turning as compared to grinding.

7. Energy consumption (i.e. power), coolant, and chip recycling will have ecological benefits in hard turning.
8. Multiple operations in turning process will offer greater flexibility to automate the process, viz. tool changing located at the turning centre.
9. Dry cutting ensures environment-friendly machining practice in turning operation. This will limit the cost of disposal, maintenance, storage, and associated health diseases.

However, the hard turning process possesses a few technical difficulties such as [34, 54]:

1. The tooling cost incurred in hard turning is comparatively higher than that of grinding.
2. Although tailstock supports the long and thin workpieces, the higher cutting pressures in hard turning induce chatter and will have determinantal effects on tool and work material. The technical limitation with regard to length-to-diameter (L/D) ratio of the workpiece material should not exceed the ratio of 4:1.
3. The accuracy (dimensional tolerance and surface finish) of machined parts in hard turning depends on the degree of machine rigidity. Higher rigidity and damping characteristics can be attained with a few additional features in the machine system, such as the machine bases made with polymer composite reinforcement, direct-seating-collected spindle arrangement, and hydrostatic guideways.
4. Rapid tool wear occurs in hard turning which deteriorates both tool life and machined surface texture.
5. The invisible white layer formation on work surface is most common in hard turning operation. The white layer thickness increases with tool wear and favours to delaminate, leading to bearing failure.

Latest research and development in the cutting-edge technology will limit most of the above-said technical difficulties of hard turning and will provide significant benefits. The factors such as high production rate, minimized machining lead time, reduced energy consumption, dry cutting condition, and easy recycling of chips will offset the cost of tooling and cost of using machine with high rigidity [55, 56]. Further, use of advanced cutting tools reduces the problem with rapid tool wear, which will help to maintain integrity on the machined surface [25, 56]. Moreover, prerequisite knowledge of tool material grade, tool geometry and holder, and cutting conditions are must to yield efficient and stable machining process [54, 56].

1.3.2 Hard Broaching

Broaching process enables production of large quantity with complex profile parts economically. The growing demand for parts with consistent quality in surface integrity and dimensional accuracies can be met by broaching process [34]. This is the only metal cutting process, which will not require the feed motion as multi-point cutting tool with the cutting edges possessing increased cutting depth from one tooth with respect to other. The tool in broaching operation will define the undeformed chip thickness [57]. Further, the geometry of tooth profile will help to remove the material with greater depth in a single stroke. Either linear or helical cutting motion can be accomplished with broaching process. This feature enables the broaching process to operate in both horizontal and vertical directions. Internal broaching operation is used to enlarge or finish the pre-machined hole. Internal broaching tool may be operated either of pushed or of pulled type in semi-finished hole. In push-type broaching operations, the parts generally experience higher compressive load, which may result in bending or buckling of parts. Hence, to avoid the said negative effect, the tool length is made shorter. Typical applications of internal broaching are manufacturing of profiled holes, namely spline, polygonal, quadrangle shape, keyways, straight teeth, helical flute of internal spur gears, etc. External broaching is used to machine parts to obtain required surface profile. It is important to note that series of rough teeth followed by semi-finish, finish, and burnishing teeth are arranged in series which will enable roughing, semi-finishing, and finishing operations in a single stroke. The external surface broaching has potential advantages in terms of productivity and quality as compared to other surface machining processes (i.e. milling, shaping, and planning). The external broaching tools may be of both push and pull types. A few applications of external surface broaching are grooves, slots, keyways, flat, peripheral and contour surfaces, external splines, and external spur gear teeth. The cutting speed in broaching will be in the range of 0.5–6 m/min. The estimated cutting speed is comparatively 10–15 times lesser than turning and milling in machining of steels and alloys [58]. In other words, the cutting speed of broaching is found to be 60 m/min whereas it is 300 m/min in case of hard turning [37]. The recent development in hard alloyed broach tool (coated and non-coated cementite carbides, PCBN) resulted in greater stability to withstand fatigue and enhanced the reliability of tool [58]. The major disadvantages of broaching process are: it is limited to machine parts having through holes and surfaces, it operates at low cutting speed, optimal design for manufacture of broaches is difficult and treated as expensive, it needs a separate broach tool whenever there is a change in geometry (shape and size) of the part, and it is economical for only large-scale production of parts [57, 59]. In addition, while machining harder materials with higher cutting speed, the tool has to function under higher temperature and results in a rapid tool wear. Hence, under these circumstances broaching tool will malfunction and result in tool wear, tool chipping, and teeth breakage, which could affect the surface quality of the parts [60]. Further, this will result in a reduced productivity.

1.3.3 Hard Boring

Boring is also closely referred to as internal turning and reaming operations, wherein it serves as the finishing of the drilled hole or increases the diameter of a hole. The main objectives of the boring and reaming operations are [34]: (1) to maintain the hole diameter with precise dimensional tolerance and accuracy by keeping the surface roughness of part to a minimum value and (2) boring operation also serves to ensure centricity, circularity, etc. The technical benefits of boring operation include machining of large diameter holes [61]. Machining of the hardened steel with hardness of 63 HRC will reduce the machining time from 1560 s, in case of internal grinding operation to 140 s in boring operation [34]. The economic benefit of machining with boring operation will result in 35% reduction in wages, machinery, and operating cost and 55% reduction in equipment cost as compared to grinding process [61]. Carbon boron nitride (CBN) tool with defined geometry is treated more economically as compared to hard turning and grinding. The limitation of boring operation is with reference to surface quality of machined part [61, 62]. CBN tool performs machining with superior quality of machining even at dry condition and is treated as environment-friendly machining at low cost [63]. The material removal phenomenon in hard boring process will result in higher plastic deformation and thermal stresses as compared to hard turning of hardened steels. This has determinantal effect on tool wear, hardness, alter in phase structure, and wear resistance of the tool.

1.3.4 Hard Milling

Hard milling technology has drawn a significant attention in the manufacturing of dies and moulds, which may partially (i.e. deep grooves and internal corners still require EDM) replace the currently employed, high-cost electric discharge machining (EDM) [34, 37]. The near-net shape manufacturing capability of hard milling to produce complex-shaped die and mould parts with minimum scrap and production time has made it more attractive [37, 64]. In machining hardened work material (i.e. above 45 HRC), milling process will result in large amount of buckling, bending, and torsional stresses, whereas only bending stress is encountered in turning [65]. The developed higher stresses will cause vibration and deflection, which leads to premature failure of the tool holder [65]. Therefore, key to successful milling of parts will involve selection of appropriate material, geometry of tool holder, which will resist bending (due to overhang length of the holder is too long) and breaking and will also result in high energy dissipation rate [65, 66]. Milling process can produce machined parts with good surface quality, which will result in a better fatigue life as compared to grinding [67]. In addition, there are many advantages such as reduced changes in microstructure phases and hardness, reduced lead time and number of necessary machine tools, economical machining, faster production rate, greater flexibility, environment-friendly, reduced compressive residual stress and minimum distortion, and improved surface integrity [68–70]. The reduction in tool life is said to be the major drawback in hard milling process [67]. The prominent disadvantages include tool wear and life and its impact on quality of surface and machinability. This drawback can be overcome with an appropriate choice of workpiece–tool material combination, tool geometry, tool holder, use of coolant, and cutting parameters [67, 71].

1.4 Challenges in Machining of Hard Materials

In ancient times, naturally available materials (soil, stone, rocks, bones, etc.) possessing higher hardness were used to cut different parts. Later, many researchers started searching for a new range of materials possessing inherent characteristics for diverse applications. This led to the discovery of new materials and modification of the existing materials to enhance some specific properties. Scientists and engineers put lot of inventions to develop the materials possessing high strength-to-weight ratio and low cost. Further, researchers have put lot of efforts to develop materials with high strength, high resistance to corrosion, low cost. The development was mainly focussed to satisfy specific functional requirements. The materials developed so far have possessed at least few of the following features:

1. High strength-to-weight ratio.
2. High stiffness and toughness.
3. High fatigue resistance.
4. High hot hardness.
5. Better thermal conductivity and heat capacity.
6. Oxidation and corrosion resistance.

1.4.1 Steels

Steel is a family of materials, which include iron and carbon as major constituents. Steels are widely used in buildings, structures, machines, tools, marine applications, automobile, aerospace, and so on due to its inherent properties, such as high tensile strength, modulus of elasticity, fracture toughness, fatigue resistance, and low cost [72–74]. Moreover, the properties can be altered and controlled through heat treatment. In general, steels are processed through heat treatment wherein the phases of materials will improve the hardness, strengths, and wear resistance. Alternatively, the properties are also improved with other treatment processes such as cryogenic with heat treatment, case hardening, flame hardening, gas nitriding, carburizing, and tempering [75, 76]. The aforementioned process improve the material strength, and hardness which offers greater resistance to deform, and results in machining a difficult process.

1.4.2 Titanium and Its Alloys

Titanium and its alloy are attractive functional materials due to their unique features, such as high strength-to-weight ratio, stiffness, ability to retain hardness at elevated temperature, toughness, high corrosion, and fatigue resistance. These inherent properties made these materials more popular in many industrial and commercial applications. A few applications include petroleum and oil refineries, nuclear and food processing appliances, chemical, missile, biomedical-based surgical implantation, electronics, and marine applications [77, 78]. Machining of titanium and its alloys is classified as difficult-to-cut materials due to their chemical reactivity and poor thermal conductivity [77, 79]. This chemical affinity will result in removing some important alloying elements from cutting tool. Further, welding of some chip material will result in chipping of tool and early tool failure [78]. Low material thermal conductivity will (i.e. 1/6th of steel) generally increase the temperature at tool–work interface and adversely affect the tool life [78].

1.4.3 Super-Alloys

Super-alloys are known for their heat-resistant properties along with high strength, stiffness, and toughness. Nickle-, iron-, and cobalt-based alloys make the family of super-alloys and are known for their high strength and hardness at elevated temperatures, low chemical affinity, and low thermal diffusivity [80, 81]. Properties of super-alloy made them suitable for the applications, where the parts are subjected to high temperature. The applications include gas turbine, blades, aviation, medical implants and automotive, marine, metal processing dies, chemical and petrochemical processing machineries [82–84]. The super-alloy material properties will affect the machinability and inhibit their extensive use with a wide range of applications. The challenges in machining of super-alloys are explained as follows [50, 80]: (a) higher strengths are observed during machining even at elevated temperatures, (b) during machining the material undergoes work hardening, which will result in notch wear at tool nose, (c) in super-alloy the presence of abrasive carbide results in high abrasive wear which will affect the cutting tool life, (d) as a result of high cutting temperatures and chemical degradation diffusion wear might occur with the use of commercially available cutting tools, (e) adhesion or welding occurs on the rake face of a cutting tool which results in severe notching and removal of tool materials, (f) it is difficult to maintain the continuous chips during machining of super-alloys and results in degradation of cutting tool as a result of crater formation, (g) low thermal conductivity, (h) high dynamic shear strength. The said reasons directly affect the tool life and surface integrity on the machined parts and make the machining process more difficult.

1.4.4 Composite Materials and Metal Matrix Composites

A composite is a material, composed of at least two different materials, possessing different physical and chemical properties. The composite material offers significantly better properties as compared to the parent material. It is to be noted that at least one of the constituent materials will act as reinforcement and the other one as matrix. Matrix and reinforcement are the main constituent materials in composite. The functions of matrix in composites are listed below:

1. To protect the reinforcement materials.
2. To uniformly distribute the stress in the reinforcement materials.
3. To obtain the final shape for the composite part.

The reinforcement (particles and fibres) in composite materials may possess the following objectives [85]:

1. Weight reduction being the primary goal must serve as the lightweight characteristics, to reduce energy consumption.
2. To attain the highest mechanical properties (strength and hardness) of composites.
3. To reinforce the matrix in preferential directions.
4. Ensure highest fatigue strength and thermal shock resistance at elevated temperatures.
5. Enhance corrosion and wear resistance.

The properties of composite materials (physical, chemical, and mechanical) are dependent on the type of reinforcement and matrix material and on their volume fraction [86]. The classification of composite materials is done as per the matrix material used, for example, polymer matrix composite (PMC), metal matrix composite (MMC) , and ceramic matrix composites (CMC). Poor impact resistance and low strain to failure will limit the use of polymer matrix, composite particularly for the applications requiring higher strength and toughness at elevated temperature [87]. Unlike PMCs, the MMCs do not absorb moisture and are not affected by radiation. Moreover, MMCs are good conductors of heat and electricity, weldable, and ease of fastening with other materials [88]. The composite materials with excellent strength and stiffness are currently facing a few critics with respect to non-homogeneity, anisotropy and the presence of fibre reinforcement (glass, graphite, boron, alumina, and silicon carbide) whiskers and abrasive particles, making the material more difficult to machine [86]. Note that the above-said abrasive reinforcement materials are brittle and generally harder than many of the cutting tool materials. In general, composite material containing traditional metallic materials is isotropic in nature, whereas fibre-reinforced composite materials are considered as anisotropic [89]. Machining of abrasive reinforced composite materials will result in brittle fracture in front of the tool. This metal removal phenomenon will not only damage the surface integrity of work material, but also result in a rapid tool wear [90]. The mechanical properties of MMCs are more sensitive to machined surfaces (such as fatigue, creep, and corrosion cracking) [89]. The machining of MMCs is still a challenging task, which might be due to the influence of large number of variables (work material, reinforcement material type and its orientation, cutting environment, cutting variables, tool material and its geometry, machine tool rigidity, and so on) and their complexity [91]. The control of said variables will reduce the cutting forces, power consumption, tool life, surface integrity, and so on [91]. MMCs are used in a wide range of applications, including aviation, military, and automotive sectors [85]. Machinability of metal matrix composites is the major concern in industries, which limits the commercial applications [92].

1.4.5 Ceramics

Ceramics are well-known engineering materials comprising of metallic and non-metallic elements possessing attractive features, such as strengths at elevated temperature, low density, oxidation, wear and corrosion resistance, chemical inertness, stiffness, hardness, and better electrical and thermal insulating characteristics [93]. In recent years, much of researches (silicon carbide and nitride, oxides—Al_2O_3 and MgO, and silicates—borosilicate glass, mullite, lithium, and alumino-borosilicate) have been carried out on development of ceramic materials. The natural raw materials used for the production of ceramics are silica stone, silica sand, graphite, limestone, magnesia, talc, kaolin, mineral (zirconium, titanium, and borate) [94]. The properties of ceramics, such as brittleness, higher hardness, and creep resistance, make it difficult to machine by conventional process and tend to cause cracks, fractures, and edge chipping [95]. The ceramic materials possess low toughness and do not offer resistance to crack propagation. In addition, the ceramic materials pose difficulty in machining due to high tooling cost which limits the material for a wide range of applications. However, a few engineering applications, related to ceramic materials, include engine parts, biomedical or dentist implants, transport, electrical and electronic components [50].

1.5 Industrial Applications of Machined Hard Materials

The hardness value of hard materials generally lies around 45 HRC. The hard materials, namely hardened alloy steels, super-alloys, ceramics, titanium and its alloys, etc., are widely used in the major areas of mechanical engineering applications (such as automobile and aerospace sectors) for production of transmission parts. The hard materials are used for the applications which include gears, gear hob, axles, bearings, arbores, camshafts, cardan joints, driving pinions, clutch plates, crankshafts, cast iron flanges, shaft shoulders, compressor vanes and blades, and turbine vanes [34, 50]. Machining of hard materials with technological and economic benefits is a tedious task. Gear wheel bearing surfaces are typically manufactured with the hard materials [50]. Typical applications of machining hard materials in turning process include the following operations such as plane turning, boring, facing, interrupted turn, grooving, and threading.

1.6 Cutting Tool Materials

Selection of cutting tools requires prior knowledge of tool and work material properties, work and tool geometry (rake angle, nose radius, cutting-edge angle, inclination angle), tool holders, cutting conditions (cutting speed, feed, and depth of cut), and cutting environment (dry, wet, and minimum quantity lubricant) [18]. Lack of information on the aforementioned parameters may result in poor-quality machining [96]. The cutting tool geometry is a primary concern to metal cutting industries as it directly affects chip control, production rate, tool life, direction and magnitude of cutting forces, surface integrity, and machining residual stresses [96]. Selection of cutting tool material and appropriate operating conditions play a vital role in machining. For example, ceramic tool operating under high rigid condition (i.e. free from vibration) with continuous cuts at higher cutting temperatures will perform well in terms of quality and surface integrity. However, machining at low cutting speed with interrupted cuts will have greater tendency to brittle failure due to mechanical shocks [97]. The appropriate selection of machining conditions will result in better quality machined parts at reduced cost. A few desirable characteristics of cutting tool materials are discussed below [25, 97]:

(1) Ability of tool material to retain high yield strength and hardness at elevated temperatures (i.e. hot hardness).
(2) Ability of tool material to offer better resistance to chipping characteristics (i.e. fracture toughness).
(3) Ability of material to withstand cyclic cutting forces (fatigue resistance) and absorb maximum energy before fracture (toughness).
(4) Ability of material to offer resistance to wear (both abrasion and adhesion) and enhance tool life.
(5) Material should have better thermal conductivity.
(6) Ability of material to withstand both mechanical and thermal shocks.
(7) Low chemical affinity to prevent possible chemical reaction at high temperature.
(8) Ability of material to offer resistance to oxidation and corrosion.

Manufacture of cutting tools with the aforementioned characteristics is treated as a challenging task to manufacturers. Extensive research and development took place in developing the cutting tool materials across the globe. The leading manufacturers (Kennametal, General Electric, Sumitomo, Sandvik, Krupp, Ceratizit, De Beers, ISCAR) of advanced cutting tool materials and cutting tool system (tool holder, inserts or integral cutting tools etc.) are contributing to 80% in total world market [25]. The major cutting tool materials used in machining hard materials are listed with their increased hardness such as high-speed steels, carbides (sintered and coated), ceramics (alumina and silicon nitride), and extra hard materials (polycrystalline diamond and polycrystalline cubic boron nitride).

1.6.1 High-Speed Steel (HSS)

F. W. Taylor and M. White of the Bethlehem Iron Company developed high-speed steels early in the twentieth century. They found that heat treatment of a wide range of alloy steels had not only improved the hardness and strength but also helped to operate under high cutting speeds. Further, addition of alloying elements (molybdenum, chromium, vanadium, tungsten, and cobalt) to high carbon steels had improved the properties (wear resistance, toughness, hot hardness, and so on) and was suitable for machining difficult-to-cut materials [97]. It is noteworthy that hardness of HSS can be enhanced up to 75 HRC. HSS tools are manufactured through various processes such as steelmaking, forging, rolling, and machining. The technological development in powder metallurgy has enabled to obtain desired properties with uniform structure and without carbide segregation problem. Coatings (TiN, TiAlN, TiCN) applied to high-speed steels will result in improving machinability characteristics. The HSS tool will retain their hardness up to 500 °C. AISI, SAE, and BSI classify the high-speed steels as T-type and M-type. The HSS tool has shown potential applications in rough milling, gear cutting tools, disc, knives, rolls, tapping, reaming, broaching, and drilling operations. Note that HSS tool is generally not preferred for the hard machining applications due to its low hardness [25].

1.6.2 Cemented Carbides

Cemented carbides are also referred to as sintered carbide or cermets. The cemented carbides are developed from stellites. The stellites are alloys of tungsten, chromium, and cobalt. Powder metallurgy (PM) technique is employed generally to manufacture the cemented carbide tools, wherein the powdered form of carbide (tungsten or titanium) and cobalt is initially pressed in the mould and latter sintered to a temperature about 1300–1600 °C [98]. A small quantity of tantalum, titanium, or vanadium carbides are added to enhance the specific characteristics in cemented carbides. Cobalt is used as a binder, whose proportion is varied in the range of 6–12%. It is to be noted that cobalt melts at 1493 °C and forms a soluble phase with tungsten carbide grains at 1275 °C. This will reduce the porosity to a greater extent. Increased per cent of cobalt will improve the tool toughness, whereas its hardness and strength will decrease [25]. The tool material hardness is directly proportional to binder quantity and the size of tungsten carbide. Note that hardness will improve with reduced tungsten carbide grain size and the binder content and vice versa. The grain size of tungsten carbide is usually between 600 and 2100 HV. In general, cemented carbide is highly wear resistant and seen to be brittle. This feature is generally not used in planning operation, wherein the cutting edge experiences the sudden impact stresses and causes early chipping [98]. However, cemented carbide tools are best suited to cut high strength steels which will operate under high cutting speeds.

Over the last three decades, metal cutting industries have been using thin-film hard coatings and thermal diffusion processes to enhance the tool life. Recent survey has shown that the coating technology is applied to 50% of HSS, 85% of carbide, and 40% of super-hard tools [99]. Today, a wide range of coating materials (TiN, TiAlN, TiCN, MoS_2, CrN, Al_2O_3, WC-C, Hf, Si, Zr) and methods (physical vapour deposition, chemical vapour deposition, plasma-assisted, and moderate temperature) are available to offer significant benefits (tool life, high productivity, high feed cutting) to machining technology [99, 100]. It is to be noted that carbides are excellent substrate materials for all tool coatings discussed above. The significant benefits of applying coating on cutting tools are as follows [96, 100]: enhance the surface hardness and wear resistance (crater, flank, abrasive and adhesive wear), minimize coefficient of friction, ease of chip removal, reduce cutting forces and heat generation, improve corrosion and oxidation resistance, and finally improve the surface integrity of the machined parts. The main disadvantages of coating technology are: the tools are costlier and suitable for only single applications (i.e. indexable inserts), and the grinding of tools removes the coating materials [97].

1.6.3 Ceramics

Ceramic tools were added early in the 1950s. The primary constituent material in ceramics includes fine-grained aluminium oxide (Al_2O_3) and is pressed to obtain green compact at room temperature. Further, it is sintered under high pressure and temperature. The ceramics are classified into major two types such as white ceramics (pure Al_2O_3) and black ceramics (combination of Al_2O_3, TiC, and ZrO_2) [96]. Ceramic tools offer better hardness and wear-resistant characteristics which allows the material to withstand temperature above 1500 °C. Thus, machining of materials with ceramic tools made it possible to machining of high cutting speeds without the use of coolants [25]. In addition, ceramic tool has better chemical stability and will not affect machinability adversely (i.e. no diffusion wear). The surface integrity on the machined surface finish is not affected much with ceramic tools. However, lower toughness during machining will limit the application of ceramic tools. The toughness of alumina ceramic tools can be improved with the homogeneous mixture of appropriate proportions of oxides of chromium or zirconium, titanium, and magnesium [98]. In general, ceramic tools are primarily available with three major combinations, such as alumina (Al_2O_3), silicon nitride (Si_3N_4), and Si-ALON (mixed composition of Si, Al, O, and N). Al_2O_3 offers better wear resistance which is suitable to cut the family of hardened steels; however, they are brittle in nature. Si_3N_4 is known for its toughness, which suits to cut the cast iron effectively. Si-ALON properties will depend on the proportion of Al_2O_3 and Si_3N_4. Higher proportion of Al_2O_3 in Si-ALON will result in improved hardness, and the toughness will increase with the higher proportion of Si_3N_4. Ceramic tools are often best suited for machining the major class of ferrous and super-alloy materials. Machining of soft materials (i.e. copper, brass, and aluminium) with ceramic tools could always result in the formation

of built-up edge. Recent developments in nanotechnology have resulted in reinforcement of ceramic materials, suspended uniformly, and provided maximum impact
and wear resistance. Coatings to ceramic tool materials often yield good results, but
result in higher cost. Further, weak adhesion of coating materials on ceramic substrate could limit the application of coatings on ceramic tools. Machining of hard
materials with ceramics will enhance tool life by 20 times as compared to modern
commercial tools.

1.6.4 Carbon Boron Nitride (CBN) Tools

After diamond, CBN is the second known hardest synthetic material, artificially
developed for cutting difficult-to-cut materials [101]. While machining hard materials, the tools deteriorate as they undergo higher stress, temperature, and vibration. The
melting temperature of CBN material is around 2730 °C, thus possessing excellent
oxidation resistance and retaining hardness with greater stability up to a temperature of 2000 °C [102]. CBN tools perform better than ceramics in machining soft
materials [96]. It is interesting to note that the CBN tool offers better properties
(mechanical, thermal shock resistance, low wear, and excellent thermal conductivity) as compared to carbide tools [79, 103]. The only limitation is that CBN tool
must operate at under low feed rate to limit tool wear and better tool life [104]. Polycrystalline cubic boron nitride (PCBN) material was developed to perform turning
and other machining operations (including hard materials such as steels, cast iron,
and super-alloys) [105]. PCBN materials are now finding significant attention as tool
material can be welded to high temperature alloying materials by friction stir processing [106]. PCBN tool offers approximately two times better resistance to abrasion as
compared to ceramics and ten times to that of carbide. The appropriate edge preparation for the PCBN inserts is mandate for enhancing tool life. Chamfering the edge
of a cutting tool will result in a better performance for industrial products over a
period of time. However, with the development of advanced cutting tool technology,
edge preparation provides better solutions (enhanced cutting performances and tool
life) as compared to chamfered cutting edges. PCBN tool coated with thin layer of
2–15 μm as a substrate either with chemical vapour or with physical vapour deposition results in improved tool performance [96]. In addition to improved hardness,
PCBN is chemically more stable and provides protective blanket to the tool from
thermal shock resistance. Coating technology offers better technical benefits, but the
surface integrity (surface defects, finish, and residual stresses) of the coated tools is
not improved significantly. It is interesting to note that the un-coated tools have more
chances to lead to compressive residual stresses during machining operation [107].

1.6.5 Polycrystalline Diamond (PCD)

PCD is one among the constituent group of extra hard materials. Recent development in tool technology showed PCD tools used are available with three major different grades. These grades are categorized based on the grain size (50 μm of coarse grain, 5 μm of fine grain, and 0.5 μm of ultra-fine) [25]. Coarse grain-sized tools possess high abrasion resistance, which are used in machining of high silicon aluminium, graphite, grey iron, ceramics, tungsten carbide, and Kevlar materials. High abrasion resistance and excellent tool edge sharpness are the main features of fine-graded PCD tool, wherein they are suitable to machine low–medium silicon aluminium, copper, fibreglass, carbon, wood–plywood, fibreboard, and hardwoods. The ultra-fine grain size diamond tools offer superior toughness to the cutting edge of a tool and are used in machining of plastics, wood, aluminium, and copper. It is interesting to note that coarse grain size tools possess better wear resistance, whereas superior surface finish will be obtained with the ultra-fine grain size tools. PCD tool possesses superior characteristics of diamond. The composite is sintered with a metallic binder under the influence of high temperature and pressure [108]. In addition, PCD tools can offer approximately 500 times greater wear resistance than tungsten carbide [4]. PCD tools are extremely fragile which exhibits low toughness and greater tendency to react with iron [108]. PCD tools are replacing the tools with high hardness (coated and un-coated tungsten carbide, ceramics, and natural diamond) in machining of carbon fibre and metal matrix-reinforced composites, wood, plastics, ceramics, etc [96].

1.7 Selection of Cutting Tool Materials and Geometry

Selecting an appropriate tool material will define the process efficiency in terms of degree of surface integrity, accuracy, and economics. The aforementioned tool materials are of generic class, wherein a wide variety of tool materials and cutting inserts are currently developed by distinguished tool manufacturers. In addition, for sustainable manufacturing the cutting environment (i.e. dry, wet, and minimum quantity lubricant), cutting fluid (i.e. solid, liquid, and semi-solid) application, and type of cutting fluid (MOS_2, vegetable oils, synthetic/mineral oils, and so on) will influence more on the machining performances. Note that a few tool materials are more sensitive to cutting fluids, which will have negative effect on tool life and machining performances (surface integrity, dimensional accuracy, economy, etc.). For sustainable machining, the parameters such as tool geometry and shape, tool holders, machining-related variables, and cutting fluids have shown a complex nonlinear behaviour on machining performances and make it difficult to set optimum conditions. In addition, the tool material properties such as hardness, toughness, and hot hardness characteristics will decide the tool life (refer Figs. 1.3 and 1.4). Furthermore, tool geometry (nose radius, cutting-edge shape, etc.) decides the performance of machining quality

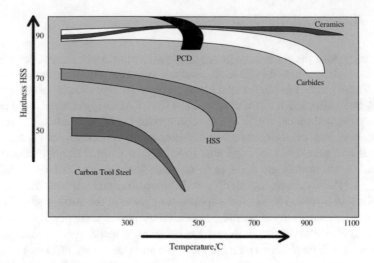

Fig. 1.3 Hardness of tool materials

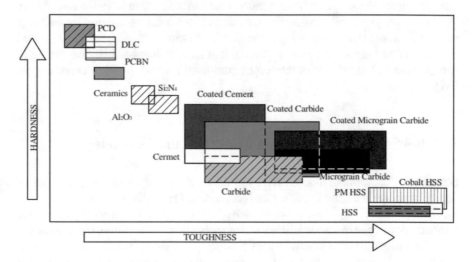

Fig. 1.4 Hardness and toughness of tool materials

(refer Fig. 1.5). Therefore, selecting an appropriate tool material and geometry for machining process is treated as a challenging task for manufacturers and requires prerequisite knowledge about machining of hard materials.

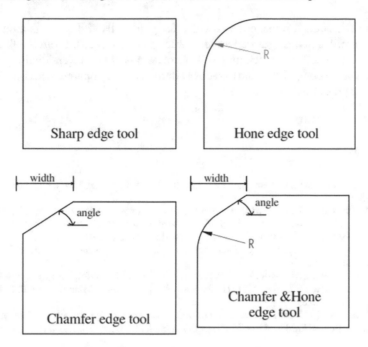

Fig. 1.5 Edge preparation for single-point cutting tools in hard turning

1.8 Advantages in Machining Hard Materials with Conventional Machining

The challenging tasks to manufacturers include continuous improvement in productivity and product quality in machining of difficult-to-cut materials. Conventional methods in hard machining yield better material removal rate and bring economy in the manufacture of machined parts. Hence, the following factors are considered in machining of hard materials to obtain best results:

(a) Rigidity of the machine tool system.
(b) Tool material properties, such as hardness and toughness.
(c) Tool geometry.
(d) Stiffness and rigidity of tool holder.
(e) Cutting environment.

The appropriate choice of said parameters will result in better tool life, material removal rate, surface integrity, dimensional accuracy, and reduced machining cost. Moreover, hard milling is a single-step process that helps to cut complex-shaped (mould and die) parts to near-net shape with less scrap rate. Hard boring operations will find major applications in cutting internal profiles, internal gear flanks, splines and polygon profiles and multiple keyways [22]. In gear manufacturing, the profile is machined with skive hobbing process, which replaces the requirement of grinding process [22]. Hard turning process will offer greater flexibility in manufacturing complex profiles in a single step as compared to grinding [22]. From the literature, it can be concluded that machining of hard materials with conventional machining process is still a competitive and attractive alternative as compared to non-traditional machining processes.

References

1. G.V. Stabler, The fundamental geometry of cutting tools. Proc. Inst. Mech. Eng. **165**(1), 14–26 (1951)
2. J.P. Davim (ed.), *Traditional Machining Processes: Research Advances* (Springer, 2014)
3. V.P. Astakhov, *Geometry of Single-Point Turning Tools and Drills: Fundamentals and Practical Applications* (Springer Science & Business Media, 2010)
4. J.P. Davim (ed.), *Machining: Fundamentals and Recent Advances* (Springer Science & Business Media, 2008)
5. R.B.D. Pereira, C.H. Lauro, L.C. Brandão, J.R. Ferreira, J.P. Davim, Tool wear in dry helical milling for hole-making in AISI H13 hardened steel. Int. J. Adv. Manuf. Technol. **101**(9–12), 2425–2439 (2018)
6. J. Vivancos, C.J. Luis, L. Costa, J.A. Ortız, Optimal machining parameters selection in high speed milling of hardened steels for injection moulds. J. Mater. Process. Technol. **155**, 1505–1512 (2004)
7. H. Coldwell, R. Woods, M. Paul, P. Koshy, R. Dewes, D. Aspinwall, Rapid machining of hardened AISI H13 and D2 moulds, dies and press tools. J. Mater. Process. Technol. **135**(2–3), 301–311 (2003)
8. S. Dilbag, P.V. Rao, Performance improvement of hard turning with solid lubricants. Int. J. Adv. Manuf. Technol. **38**(5–6), 529–535 (2008)
9. V.N. Gaitonde, S.R. Karnik, L. Figueira, J.P. Davim, Analysis of machinability during hard turning of cold work tool steel (type: AISI D2). Mater. Manuf. Processes **24**(12), 1373–1382 (2009)
10. E.M. Trent, P.K. Wright, *Metal cutting* (Butterworth-Heinemann, MA, 2000)
11. T.H.C. Childs, K. Maekawa, T. Obikawa, Y. Yamane, *Metal Machining: Theory and Applications* (Elsevier, MA, 2000)
12. A. Shokrani, V. Dhokia, S. Newman, Investigation of the effects of cryogenic machining on surface integrity in CNC end milling of Ti-6Al-4V titanium alloy. J. Manuf. Process. **21**, 172–179 (2016)
13. R.V. Rao, V.D. Kalyankar, Parameter optimization of modern machining processes using teaching–learning-based optimization algorithm. Eng. Appl. Artif. Intell. **26**(1), 524–531 (2013)
14. W. Wei, Z. Di, D.M. Allen, H.J.A. Almond, Non-traditional machining techniques for fabricating metal aerospace filters. Chin. J. Aeronaut. **21**(5), 441–447 (2008)
15. F. Cus, J. Balic, Optimization of cutting process by GA approach. Robot. Comput. Integr. Manuf. **19**(1–2), 113–121 (2003)
16. T.N. Wong, S.L. Siu, A knowledge-based approach to automated machining process selection and sequencing. Int. J. Prod. Res. **33**(12), 3465–3484 (1995)
17. I. Mukherjee, P.K. Ray, A review of optimization techniques in metal cutting processes. Comput. Ind. Eng. **50**(1–2), 15–34 (2006)
18. B. Arezoo, K. Ridgway, A.M.A. Al-Ahmari, Selection of cutting tools and conditions of machining operations using an expert system. Comput. Ind. **42**(1), 43–58 (2000)
19. R. Suresh, S. Basavarajappa, V.N. Gaitonde, G.L. Samuel, Machinability investigations on hardened AISI 4340 steel using coated carbide insert. Int. J. Refract Metal Hard Mater. **33**, 75–86 (2012)

20. H. Aouici, H. Bouchelaghem, M.A. Yallese, M. Elbah, B. Fnides, Machinability investigation in hard turning of AISI D3 cold work steel with ceramic tool using response surface methodology. Int. J. Adv. Manuf. Technol. **73**(9–12), 1775–1788 (2014)
21. G. Bartarya, S.K. Choudhury, State of the art in hard turning. Int. J. Mach. Tools Manuf **53**(1), 1–14 (2012)
22. F. Klocke, E. Brinksmeier, K. Weinert, Capability profile of hard cutting and grinding processes. CIRP Ann. Manuf. Technol. **54**(2), 22–45 (2005)
23. Y.K. Chou, Hui Song, Tool nose radius effects on finish hard turning. J. Mater. Process. Technol. **148**(2), 259–268 (2004)
24. H.K. Tonshoff, H.G. Wobker, D. Brandt, Hard turning—Influence on the workpiece properties. Trans. North Am. Manuf. Res. Inst. SME **23**, 215–220 (1995)
25. L.N.L. De Lacalle, A. Lamikiz, J.F. de Larrinoa, I. Azkona, Advanced cutting tools, in *Machining of Hard Materials* (Springer, London, 2011), pp. 33–86
26. E. Kuram, B. Ozcelik, E. Demirbas, Environmentally friendly machining: vegetable based cutting fluids, in *Green Manufacturing Processes and Systems* (Springer, Berlin, Heidelberg, 2013), pp. 23–47
27. W.W. Badiuzaman, M.A. Karim, N.A. Derahman, N.M. Amran, M.M. Isa, Analysation of performances of CNC high speed milling machine using multi-walled carbon nanotubes as additives in cutting fluid. Materialwiss Werkstofftechnik **49**(4), 494–499 (2018)
28. G. Byrne, E. Scholta, Environmentally clean machining processes—a strategic approach. CIRP Ann. Manuf. Technol. **42**(1), 471–474 (1993)
29. Y.M. Shashidhara, S.R. Jayaram, Vegetable oils as a potential cutting fluid—an evolution. Tribol. Int. **43**(5–6), 1073–1081 (2010)
30. P.S. Sreejith, B.K.A. Ngoi, Dry machining: machining of the future. J. Mater. Process. Technol. **101**(1–3), 287–291 (2000)
31. A.E. Diniz, R. Micaroni, Cutting conditions for finish turning process aiming: the use of dry cutting. Int. J. Mach. Tools Manuf. **42**(8), 899–904 (2002)
32. R.W. Cumberland, M.B. Weinberger, J.J. Gilman, S.M. Clark, S.H. Tolbert, R.B. Kaner, Osmium diboride, an ultra-incompressible, hard material. J. Am. Chem. Soc. **127**(20), 7264–7265 (2005)
33. S. Singh, S. Maheshwari, P.C. Pandey, Some investigations into the electric discharge machining of hardened tool steel using different electrode materials. J. Mater. Process. Technol. **149**(1–3), 272–277 (2004)
34. V.P. Astakhov, Machining of hard materials–definitions and industrial applications, in *Machining of Hard Materials* (Springer, London, 2011), pp. 1–32
35. M.W. Cook, P.K. Bossom, Trends and recent developments in the material manufacture and cutting tool application of polycrystalline diamond and polycrystalline cubic boron nitride. Int. J. Refract Metal Hard Mater. **18**(2–3), 147–152 (2000)
36. D. Umbrello, J. Hua, R. Shivpuri, Hardness-based flow stress and fracture models for numerical simulation of hard machining AISI 52100 bearing steel. Mater. Sci. Eng. A **374**(1–2), 90–100 (2004)
37. G. Grzesik, Machining of hard materials, in *Machining: Fundamentals and Recent Advances*, ed. by P. Davim (Springer, London, 2008), pp. 97–126
38. S. Malkin, C. Guo, *Grinding Technology: Theory and Application of Machining with Abrasives* (Industrial Press Inc., 2008)
39. S. Jha, V.K. Jain, Nanofinishing techniques, in *Micromanufacturing and Nanotechnology* (Springer, Berlin, Heidelberg, 2006), pp. 171–195
40. E.O. Ezugwu, Key improvements in the machining of difficult-to-cut aerospace superalloys. Int. J. Mach. Tools Manuf. **45**(12–13), 1353–1367 (2005)
41. J. Rech, A. Moisan, Surface integrity in finish hard turning of case hardened steel. Int. J. Mach. Tools Manuf. **43**(5), 543–550 (2003)
42. D.I. Lalwani, N.K. Mehta, P.K. Jain, Experimental investigations of cutting parameters influence on cutting forces and surface roughness in finish hard turning of MDN250 steel. J. Mater. Process. Technol. **206**(1–3), 167–179 (2008)

43. G. Byrne, D. Dornfeld, B. Denkena, Advancing cutting technology. Ann CIRP **52**(2), 483–507 (2003)
44. D. Singh, P.V. Rao, A surface roughness prediction model for hard turning process. Int. J. Adv. Manuf. Technol. **32**(11–12), 1115–1124 (2007)
45. W. König, R. Komanduri, H.K. Toenshoff, G. Ackershott, Machining of hard materials. CIRP Ann. **33**(2), 417–427 (1984)
46. H.K. Tönshoff, F. Kroos, C. Marzenell, High-pressure water peening-a new mechanical surface-strengthening process. CIRP Ann. **46**(1), 113–116 (1997)
47. A. Das, S.K. Patel, T.K. Hotta, B.B. Biswal, Statistical analysis of different machining characteristics of EN-24 alloy steel during dry hard turning with multilayer coated cermet inserts. Measurement **134**, 123–141 (2019)
48. M.A. Sampaio, Á.R. Machado, C.A.H. Laurindo, R.D. Torres, F.L. Amorim, Influence of minimum quantity of lubrication (MQL) when turning hardened SAE 1045 steel: a comparison with dry machining. Int. J. Adv. Manuf. Technol. **98**(1–4), 959–968 (2018)
49. A. Alok, M. Das, Multi-objective optimization of cutting parameters during sustainable dry hard turning of AISI 52100 steel with newly develop HSN2-coated carbide insert. Measurement **133**, 288–302 (2019)
50. H.A. Kishawy, A. Hosseini, Machining difficult-to-cut materials. Mater. Form. Mach. Tribol. (2019). https://doi.org/10.1007/978-3-319-95966-5_4
51. R. Kumar, A.K. Sahoo, P.C. Mishra, R.K. Das, Measurement and machinability study under environmentally conscious spray impingement cooling assisted machining. Measurement **135**, 913–927 (2019)
52. S. Debnath, M.M. Reddy, A. Pramanik, Dry and near-dry machining techniques for green manufacturing, in *Innovations in Manufacturing for Sustainability. Materials Forming, Machining and Tribology*, ed. by K. Gupta (Springer, Cham, 2019)
53. B.P. Erdel, *High-Speed Machining. Society of Manufacturing Engineers* (2003)
54. R. Suresh, S. Basavarajappa, V.N. Gaitonde, G.L. Samuel, J.P. Davim, State-of-the-art research in machinability of hardened steels. Proc. Inst. Mech. Eng. Part B: J. Eng. Manuf. **227**(2), 191–209 (2013)
55. R. Hasan, Why are you still grinding? Manuf. Eng. (USA) **120**(2), 76 (1998)
56. H. Tonshoff, C. Arendt, R. Ben Amor, Cutting of hardened steel. Ann. CIRP **49**, 547–566 (2000)
57. H.K. Toenshoff, B. Denkena, *Basics of Cutting and Abrasive Processes*. Lecture Notes in Production Engineering. https://doi.org/10.1007/978-3-642-33257-9_11 (2013)
58. V.F. Makarov, D.I. Tokarev, V.R. Tyktamishev, High speed broaching of hard machining materials. Int. J. Mater. Form. **1**(1), 547–550 (2008)
59. U. Kokturk, E. Budak, Optimization of broaching tool design, in *Proceeding of the CIRP ICME*, 4 (2004)
60. D. Shi, D.A. Axinte, N.N. Gindy, Development of an online machining process monitoring system: a case study of the broaching process. Int. J. Adv. Manuf. Technol. **34**(1–2), 34–46 (2007)
61. J. Kundrák, A.G. Mamalis, A. Markopoulos, Finishing of hardened boreholes: grinding or hard cutting? Mater. Manuf. Processes **19**(6), 979–993 (2004)
62. Y. Matsumoto, F. Hashimoto, G. Lahoti, Surface integrity generated by precision hard turning. CIRP Ann. Manuf. Technol. **48**(1), 59–62 (1999)
63. J. Kundrák, Hard boring of gears. J. Prod. Process. Syst. **6**(1), 61–70 (2012)
64. W. Li, Y. Guo, C. Guo, Superior surface integrity by sustainable dry hard milling and impact on fatigue. CIRP Ann. Manuf. Technol. **62**(1), 567–570 (2013)
65. H. Çalışkan, C. Kurbanoğlu, P. Panjan, D. Kramar, Investigation of the performance of carbide cutting tools with hard coatings in hard milling based on the response surface methodology. Int. J. Adv. Manuf. Technol. **66**(5–8), 883–893 (2013)
66. P. Chatterjee, S. Chakraborty, Material selection using preferential ranking methods. Mater. Des. **35**, 384–393 (2012)

67. A. Iqbal, H. Ning, I. Khan, L. Liang, N.U. Dar, Modeling the effects of cutting parameters in MQL-employed finish hard-milling process using D-optimal method. J. Mater. Process. Technol. **199**(1–3), 379–390 (2008)

68. D.A. Axinte, R.C. Dewes, Surface integrity of hot work tool steel after high speed milling-experimental data and empirical models. J. Mater. Process. Technol. **127**(3), 325–335 (2002)

69. S. Zhang, Y.B. Guo, Taguchi method based process space for optimal surface topography by finish hard milling. J. Manuf. Sci. Eng. **131**(5), 051003 (2009)

70. D.W. Wu, Y. Matsumoto, The effect of hardness on residual stresses in orthogonal machining of AISI 4340 steel. J. Eng. Ind. **112**(3), 245–252 (1990)

71. H. Çalışkan, B. Kurşuncu, C. Kurbanoğlu, Ş.Y. Güven, Material selection for the tool holder working under hard milling conditions using different multi criteria decision making methods. Mater. Des. **45**, 473–479 (2013)

72. O. Grässel, L. Krüger, G. Frommeyer, L.W. Meyer, High strength Fe–Mn–(Al, Si) TRIP/TWIP steels development—properties—application. Int. J. Plast **16**(10–11), 1391–1409 (2000)

73. R.L. Klueh, Elevated temperature ferritic and martensitic steels and their application to future nuclear reactors. Int. Mater. Rev. **50**(5), 287–310 (2005)

74. M. Finšgar, J. Jackson, Application of corrosion inhibitors for steels in acidic media for the oil and gas industry: a review. Corros. Sci. **86**, 17–41 (2014)

75. J.R. Davis, *Surface Hardening of Steels* (ASM International, Materials Park, OH, 2002), p. 227

76. K. Moore, D.N. Collins, Cryogenic treatment of three heat-treated tool steels, in *Key Engineering Materials*, vol. 86 (1993), pp. 47–54

77. A.R. Machado, J. Wallbank, Machining of titanium and its alloys—a review. Proc. Inst. Mech. Eng. Part B J. Eng. Manuf. **204**(1), 53–60 (1990)

78. E.O. Ezugwu, Z.M. Wang, Titanium alloys and their machinability—a review. J. Mater. Process. Technol. **68**(3), 262–274 (1997)

79. X. Yang, C. Richard Liu, Machining titanium and its alloys. Mach. Sci. Technol. **3**(1), 107–139 (1999)

80. E.O. Ezugwu, Z.M. Wang, A.R. Machado, The machinability of nickel-based alloys: a review. J. Mater. Process. Technol. **86**(1–3), 1–16 (1999)

81. I.A. Choudhury, M.A. El-Baradie, Machinability of nickel-base super alloys: a general review. J. Mater. Process. Technol. **77**(1–3), 278–284 (1998)

82. R.R. Boyer, Attributes, characteristics, and applications of titanium and its alloys. J. Mater. **62**(5), 21–24 (2010)

83. C.N. Elias, J.H.C. Lima, R. Valiev, M.A. Meyers, Biomedical applications of titanium and its alloys. J. Mater. **60**(3), 46–49 (2008)

84. I.V. Gorynin, Titanium alloys for marine application. Mater. Sci. Eng. A **263**(2), 112–116 (1999)

85. U.K. Karl, Metal matrix composites: custom-made materials for automotive and aerospace engineering, in *Basics of Metal Matrix Composites*, ed. by K.U. Kainer (Wiley-VCH Verlag GmbH & Co. KGaA, Weinheim, FRG, 2006)

86. R. Teti, Machining of composite materials. CIRP Ann. Manuf. Technol. **51**(2), 611–634 (2002)

87. T.W. Chou, J.M. Yang, Structure-performance maps of polymeric, metal, and ceramic matrix composites. Metall. Trans. A **17**(9), 1547–1559 (1986)

88. A. Evans, C. San Marchi, A. Mortensen, *Metal Matrix Composites in Industry: An Introduction and a Survey* (Springer Science & Business Media, 2013)

89. A.B. Sadat, Surface integrity when machining metal matrix composites, in *Machining of Metal Matrix Composites* (Springer, London, 2012), pp. 51–61

90. C.R. Dandekar, Y.C. Shin, Modeling of machining of composite materials: a review. Int. J. Mach. Tools Manuf. **57**, 102–121 (2012)

91. N. Muthukrishnan, M. Murugan, K.P. Rao, Machinability issues in turning of Al-SiC (10p) metal matrix composites. Int. J. Adv. Manuf. Technol. **39**(3–4), 211–218 (2008)

92. I.A. Di, A. Paoletti, Machinability aspects of metal matrix composites, in *Machining of Metal Matrix Composites* (Springer, London, 2012), pp. 63–77

93. X.Q. Cao, R. Vassen, D. Stoever, Ceramic materials for thermal barrier coatings. J. Eur. Ceram. Soc. **24**(1), 1–10 (2004)
94. Y. Imanaka, Y. Suzuki, T. Suzuki, K. Hirao, T. Tsuchiya, H. Nagata, J.S. Cross, *Advanced Ceramic Technologies and Products* (The Ceramic Society of Japan, 2012). https://doi.org/10.1007/978-4-431-54108-0_2
95. A. Gorin, M.M. Reddy, Advanced ceramics: Some challenges and solutions in machining by conventional methods. Appl. Mech. Mater. **624**, 42–47 (2014)
96. V.P. Astakhov, J.P. Davim, Tools (geometry and material) and tool wear, in *Machining* (Springer, London, 2008), pp. 29–57
97. B. Mills, *Machinability of Engineering Materials* (Springer Science & Business Media, 2012). https://doi.org/10.1007/978-94-009-6631-4
98. H. Tschätsch, *Applied Machining Technology* (Springer Science & Business Media, 2010). https://doi.org/10.1007/978-3-642-01007-1
99. V.P. Astakhov, *Tribology of Metal Cutting*, vol. 52 (Elsevier, Amsterdam, 2006)
100. K.D. Bouzakis, N. Michailidis, G. Skordaris, E. Bouzakis, D. Biermann, R. M'Saoubi, Cutting with coated tools: Coating technologies, characterization methods and performance optimization. CIRP Ann. Manuf. Technol. **61**(2), 703–723 (2012)
101. Z.C. Lin, D.Y. Chen, A study of cutting with a CBN tool. J. Mater. Process. Technol. **49**(1–2), 149–164 (1995)
102. F. Klocke, *Manufacturing Processes* (Springer, Berlin, 2011)
103. Z.G. Wang, M. Rahman, Y.S. Wong, Tool wear characteristics of binderless CBN tools used in high-speed milling of titanium alloys. Wear **258**(5–6), 752–758 (2005)
104. Z.A. Zoya, R. Krishnamurthy, The performance of CBN tools in the machining of titanium alloys. J. Mater. Process. Technol. **100**(1–3), 80–86 (2000)
105. C.B. Fuller, Friction stir tooling: tool materials and designs. Frict. Stir Weld. Process. (2007), pp. 7–36
106. M.K. Besharati-Givi, P. Asadi, *Advances in Friction-Stir Welding and Processing* (Elsevier, 2014)
107. R.M. Arunachalam, M.A. Mannan, A.C. Spowage, Surface integrity when machining age hardened Inconel 718 with coated carbide cutting tools. Int. J. Mach. Tools Manuf. **44**(14), 1481–1491 (2004)
108. A. Hosseini, H.A. Kishawy, Cutting tool materials and tool wear, in *Machining of Titanium Alloys* (Springer, Berlin, Heidelberg, 2014), pp. 31–56

Chapter 2
Studies on Machining of Hard Materials

Over the years, machining industries are continuously striving to manufacture the parts at reduced cost and improved quality. This can be achieved by selecting appropriate set of tool–work materials and effective modelling and optimization of the process. Optimized grades of high-speed steel (HSS) are used to be treated as ultimate tool material till the 1930s [1]. However, American metalworking industry had shown three-time improvement in productivity with the use of same machines and manpower during the period 1939–1945. These revolutionary changes in machining were attributed to the development or invention of cemented carbide tools. The cemented carbide tools, possessing superior properties (improved tool life, ability to cut at faster rate, high hot strength, and so on), reduced the manufacturing costs of various machining processes (i.e. turning, milling, drilling). In addition to an appropriate choice of tool–workpiece material, the improvement in product quality and processes directs the economic growth of industries [2]. Many research efforts were made to improve the process/product quality with the use of traditional try-error method, classical engineering experiments, and analytical, numerical, and optimization approaches [3–11]. The above methods were employed by distinguished researchers focussed mainly on establishing input–output relationships in machining with an objective to optimize the product quality and cost. In view of the above, the present chapter discusses various research works carried out in hard turning process.

2.1 Hard Turning Process

Hard turning is a process of machining metals with hardness greater than 45 HRC, by employing a single-point cutting tool on the lathe or turning centre. Chapter 1 describes the process capabilities, technical difficulties, and applications of hard turning process. Surface finish of turned parts is still challenging, although the process has proved significant technological and economic benefits over grinding

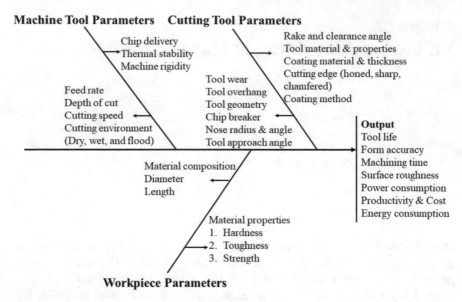

Fig. 2.1 Ishikawa diagram for hard turning process

process [12–15]. Hence, there is tremendous requirement to investigate and analyse the factors influencing machining process. Thereby, this chapter presents an overview of previous research work carried out on turning of hard metals. The majority of research work discussed involves influencing variables (tool and work material properties, cutting environment, machining conditions, and so on) and performance quality characteristics (i.e. machinability parameters such as surface finish, tool wear and life, cutting forces, material removal rate, surface integrity, and so on) in hard turning. The influence of variables on the machinability of hard turning process is represented in Fig. 2.1 as Ishikawa diagram.

In general, engineers and researchers conduct experiments for the following reasons: (a) to determine the performance (i.e. output) at a single point (i.e. single input variable) in a design space, (b) to understand and learn about the entire design space (i.e. to determine the impact of all combination of design or influencing variables present in the system), (c) to establish or develop models that relate process of input–output relationships, (d) to attain continuous improvements in the performance of the process or design. In view of above, several research efforts are made across the globe in developing techniques or approaches to study the variables that improve the machinability performances. Hence, it is important to know these techniques in detail.

2.2   Classical Engineering Experimental Approach or One Factor at a Time (OFAT)

Engineers and scientists conduct experimental studies utilizing one-factor-at-a time technique, to understand the influence of variable on performance. In this approach, one variable or factor is varied at a time and keeping the rest of factors at fixed value. The procedure is repeated depending on the number of factors to be studied. The experimenter/investigator gain information about one variable during experimental run. Table 2.1 presents the summary of cutting parameters and the performance characteristics using OFAT approach.

Significant advantages of one-factor-at-a-time approach are listed below [16–18]:

(a)   It concentrates or estimates the design space.
(b)   It provides information for the experimenter wishes to respond quickly to data and decide operating range (i.e. maximum and minimum value) for a factor.
(c)   It helps to know the sensitivity of the independent factor that helps to screen the insignificant factors and rank the factors.

There are many limitations of one-factor-at-a-time approach discussed in the literature, such as [19–24]:

(a)   It requires many experimental trials to gain the same degree of precision in estimating factor effects and optimize process.
(b)   It does not estimate the factor interaction.
(c)   The conclusions drawn from the analysis may not be made general.
(d)   The recommended optimal setting may not yield global solution always.
(e)   The methodology does not establish predictive equations, which are used in modelling and optimization.
(f)   This method is not suitable to study interaction factors and establish nonlinear relationships with responses. Thus, it limits to attain the local knowledge information about a process.

2.3   Numerical Modelling Approach

Practical difficulty in obtaining the temperature, force and tool wear data during experimentation limits the researchers to develop more precise and reliable tools based on experiments. Numerical methods based on finite element analysis (FEA) provide an effective path to investigate the factors influencing the hard turning process. Many efforts are made in this direction to utilize the numerical simulation techniques (i.e. FEA and ABAQUS) to study the mechanics and dynamics involved in the turning process. FEA model was used in three different coatings, applied on the cutting tool (i.e. PCBN), and their behaviour was observed on tool wear, cutting force, and surface finish while machining AISI 4340 hardened steel [25]. Two-dimensional thermomechanical integrated FE model was used to simulate the SKD11

Table 2.1 Outline of cutting parameters and performance characteristics studied using OFAT

Authors	Cutting parameters				Performance characteristics
	Work material and hardness (HRC)	Tool material	CS (m/min), FR (mm/rev)	DOC (mm), NR (mm)	CF (N), TW (mm), SR (μm)
[42]	AISI 4340	CBN-L + TiN-coated ceramic	CS: 100, 125, 150	DOC: 0.25	TW: 0.2
	53 HRC		FR: 0.1, 0.15, 0.2	NR: 0.8	SR: 0.3–1.3
[43]	AISI D2	PCBN	CS: 70, 95, 120	DOC: 0.5	TW: 0.3
	62 HRC		FR: 0.08, 0.14, 0.2	NR: 0.8	SR: 0.05–1.0
[44]	AISI 52100	CBN-H	CS: 60, 120, 240	DOC: 0.01-0.25	TW: 0.25
	63 HRC		FR: 0.0125	NR: 0.8	SR: 0.4
[45]	AISI 52100	CBN	CS: 91, 145, 183	DOC: 0.2–0.5	TW: 0.15–0.2
	62 HRC		FR: 0.07–0.15	NR: 0.8	SR: <0.3
[46]	AISI O1	PCBN	CS: 125	DOC: 0.15	SR: <0.2
	56 ± 2 HRC		FR: 0.15	NR: 0.8	
[47]	AISI D2	(Al$_2$O$_3$/TiCN) Ceramic	CS: 100, 140, 200	DOC: 0.4	TW: 0.3
	60 HRC		FR: 0.06	NR: 0.8	SR: 0.2
[48]	AISI D2	PCBN insert	CS: 140, 355, 500	DOC: 0.2–0.6	SR: 0.3–0.7
	62 HRC		FR: 0.05–0.2	NR: 0.4, 0.8, 1.2	
[49]	AISI 52100	PCBN	CS: 120	DOC: 0.1–0.2	SR: 0.2–0.3
	58–62 HRC		FR: 0.1–0.2	NR: 0.4, 0.8, 1.2	
[50]	AISI 4140	TiAlN PVD-coated carbide	CS: 210–410	DOC: 1.0	TW: 0.2
	60 HRC		FR: 0.14	NR: 0.8	SR: 0.5
[51]	AISI 4340	CBN	CS: 150	DOC: 0.15	TW: 0.2
	56 HRC		FR: 0.08		SR: <0.2
[52]	AISI H13	CBN	CS: 100, 200	DOC: 0.25	SR: 0.25–0.35
	50–55 HRC		FR: 0.05, 0.1, 0.2	NR: 10 μm (honed)	
[53]	AISI D2	Ceramic (wiper insert)	CS: 80, 115, 150	DOC: 0.2	CF: 154–350
					TW: 0.04–0.15

(continued)

Table 2.1 (continued)

Authors	Cutting parameters				Performance characteristics
	Work material and hardness (HRC)	Tool material	CS (m/min), FR (mm/rev)	DOC (mm), NR (mm)	CF (N), TW (mm), SR (μm)
	60 ± 1 HRC		FR: 0.05, 0.1, 0.15	NR: 10 μm (honed)	SR: 0.2–0.7
[54]	AISI 5115	CBN (WBN560)	CS: 150	DOC: 0.1–0.2	TW: <0.2
	60–62 HRC		FR: 0.1	NR: 0.1–1.2	
[55]	AISI-D3	CBN	CS: 70, 100, 130	DOC: 0.1	TW: 0.3
	60 HRC		FR: 0.05		SR: <0.1
[56]	AISI 52100	Al₂O₃–TiCN	CS: 100–300	DOC: 0.5	SR: 0.1–0.9
	63 HRC		FR: 0.07–0.14	NR: 0.8	
[57]	AISI 4340	CBN	CS: 90	DOC: 0.1	SR: 0.23
	69 HRC		FR: 0.08	NR: 0.8	
[58]	AISI 4340	Coated carbide	CS: 150	DOC: 0.4	TW: 0.3
	47 ± 1 HRC		FR: 0.15	NR: 0.8	SR: 0.3–1.2
[59]	AISI D2	PCBN and CVD coat tungsten carbide	CS: 80, 240	DOC: 0.2	TW: 0.25
	52 HRC		FR: 0.05	NR: 0.8	
[60]	AISI 4340	Cemented carbide coated with cobalt	CS: 300–600	DOC: 0.1–0.3	TW: 0.25
	47 HRC		FR: 0.1–0.3		SR: 1.5–1.7
[61]	51 CrV₄	CBN	CS: 90–140	DOC: 0.06–0.12	CF: 200–700
	68 HRC		FR: 0.12	NR: 0.8	
[62]	AISI D2	Coated carbide	CS: 135–325	DOC: 0.2–0.6	SR: 0.2–1.8
	66 HRC		FR: 0.05–0.159		

American Iron and Steel Institute (*AISI*); carbon boron nitride (*CBN*); chemical vapour deposition (*CVD*); cutting force (*CF*) in N; cutting speed (*CS*) in m/min; depth of cut (*DOC*) in mm; feed rate (*FR*) in mm/rev; hardness Rockwell C (*HRC*); nose radius (*NR*) in mm; polycrystalline cubic boron nitride (*PCBN*); surface roughness (*SR*) in μm; tool wear (*TW*) in mm

grade steel [26]. FE model was developed to predict the tool wear of the cemented carbide tool, while machining nickel-based alloys [27]. Three-dimensional finite element models were used to study the flank wear behaviour of the tools (ceramic and ultra-fine-grained ceramic) subjected to hard turning process [28]. The commercial DEFORM 3D software platform with finite element codes was used to simulate the hard turning process [28–32]. Numerical formulation of the problem and chip formation mechanism were the two major constituent modules used in hard turning process. Johnson–Cook (J-C) material model was widely employed to describe the work material behaviour (i.e. plasticity) in machining. This model enabled the practice engineer or investigator to know the influence of various process parameters on cutting forces, temperature, chip shape and chip flow direction, friction coefficient, tool wear and life [29]. FEM was employed by various researchers to simulate the hard turning process, with a focus on work material residual stress and strain fields [33], crater wear [34], and temperature at tool–work interface [35]. Two-dimensional ABAQUS model was developed to analyse hard turning of AISI 1045 steel, using carbide cutting tools [36]. Tool and work material properties (both mechanical and thermal), tool geometry, and process boundary conditions are the prerequisite in model building/simulation.

The numerical modelling tools suffer the major weakness such as [36, 37]:

(1) Machining happens under the assumptions of plain strain condition. However, in actual practice, the cutting width is found to be more than that of the un-deformed chip thickness.
(2) In simulation, the cutting tool is assumed to behave perfectly elastic, whereas actually the elasticity of tool is comparatively larger than the workpiece. However, the said elastic deformation of a tool is treated as negligible.
(3) To limit the complexity involved in simulation, the cutting tool edge is assumed to be perfectly sharp. However, in actual turning experiments there exists a gradual loss in the sharpness of cutting edge.
(4) In simulation, the tool–work interface is modelled with certain assumptions on Coulomb friction.
(5) The numerical modelling approach may not derive generalized predictive equation and further requires expertise to interpret the results. Further, the cost of DEFORM 3D software is very high.
(6) The simulation software is restricted to test only the individual factor effects. However, experimental results showed that there exist strong interactions among the factors in hard turning process [38, 39].

2.4   Input–Output and In-Process Parameter Relationship Modelling

Modelling refers to developing the relationship between inputs and outputs of a process. Establishing in-process input–output functional relationships is the prerequisite for process optimization. Mechanistic and empirical models are generally used to develop an explicit mathematical model which are governed by the influencing process parameters [40]. Mechanistic models work with the principles of analytical or numerical concepts, which use many assumptions while developing in-process input–output relationship. In addition, mechanistic models do not provide the detailed insight of physics of the process (that is, influence of material properties, cutting environment, and geometry on cutting mechanics) [41]. Thereby, empirical models are most preferred choice over numerical modelling as they use the results of practical experiments. A few empirical models developed in-process input–output relationship for a hard turning process, and the same is presented in Table 2.2.

Many research efforts are made to develop the modelling tools, which provide satisfactory results. However, they work with different constraints and limitations which will restrict their practical utility. Therefore, an appropriate choice of suitable tools requires the detailed insight of strengths and weakness of a model. In view of the above, the well-known empirical models, used to establish input–output relationships, are discussed below.

2.4.1 Taguchi Method

Genichi Taguchi developed the robust design concepts that could improve productivity, reliability, and manufacturability of a process [91]. Taguchi technique is a systematic methodology practised in robust design, which uses planned experimental data to make meaningful analysis and draw objective conclusions [92]. In 1980, application of Taguchi method benefited many US manufacturers, namely American Telephone and Telegraph, Ford Motor, and Xerox with regard to product and process quality improvement [93]. Taguchi method uses well-defined flexible orthogonal array design, which enables the investigator to know the factor contributions on the target outputs for quality improvement with minimum experimental trials, resources, and cost [94, 95]. Thereby, Taguchi method has drawn much attention of practitioners, researchers, and engineers to apply this technique in resolving problems associated with many metal cutting processes. Taguchi method was used to limit the computational complexity and time required in simulation, viz. FEA and ABAQUS [36]. The cutting parameters (cutting speed, feed rate, rake angle, and depth of cut) were varied intentionally, and their behaviour was observed on cutting force, temperature with different cutting tool materials. Table 2.2 presents the application of Taguchi method in determining the parameters and their operating levels employed to attain optimal variable setting responsible for minimum surface finish, tool wear, and cutting forces. It is to be noted that the data presented in Table 2.2 is collected from the literature. Apart from numerous advantages, a few researchers are still criticizing the Taguchi applications due to the following reasons [2, 96]:

1. Taguchi method enables limited orthogonal arrays and thus limits to test all potential interactions of a process.
2. Taguchi method many time results in a suboptimal solution, as it is focussed mainly on reducing experimental numbers and associated cost.
3. Taguchi method prompts process optimization without the empirical or mechanistic models while performing experiments. Thus, it limits to provide the detailed knowledge or insight of parameters required to monitor or control the process.

Table 2.2 Outline of cutting parameters and performance characteristics studied using statistical methods

Authors	Cutting parameters					Performance characteristics
	Work material and hardness (HRC)	Methods (experiment and optimization)	Tool material	CS (m/min), FR (mm/rev)	DOC (mm), NR (mm)	CF (N), TW (mm), SR (μm), RS (MPa), MRR (mm^3/min)
[63]	AISI 4142	Taguchi	PCBN	CS: 100, 130, 160	DOC: 0.1, 0.35, 0.6	SR: 0.275–0.792
	55 HRC			FR: 0.05, 0.15, 0.25	NR: 0.8	
[64]	AISI 4340	Taguchi	Carbide (TiN, TiC, Al_2O_3)	CS: 140, 200, 260	DOC: 0.6, 0.8, 1.0	CF: 440–754 TW: 0.67–0.222 SR: 0.68–1.42
	48 HRC			FR: 0.1, 0.18, 0.26	NR: 0.8	
[65]	AISI H11	BBD	CBN 7020	CS: 120, 180, 240	DOC: 0.1, 0.35, 0.6	SR: 0.33–0.85
	40, 45, 50 HRC			FR: 0.08, 0.12, 0.16	NR: 0.8	
[66]	X38CrMoV5-1	FFD	CBN 7020	CS: 120, 180, 240	DOC: 0.15, 0.3, 0.45	CF: 38–396
	50 HRC	DFA		FR: 0.08, 0.12, 0.16	NR: 0.8	
[67]	42CrMo4	Taguchi	CC650 (Al_2O_3/TiC) ceramic	CS: 90, 120, 180	DOC: 0.15, 0.3, 0.45	SR: 0.2–0.8
	56 HRC	DFA		FR: 0.08, 0.12, 0.16	NR: 0.8	
[68]	AISI 4140	Taguchi	CC6050	CS: 80, 115, 150	DOC: 0.1, 0.2, 0.3	SR: 0.3–0.8
	60 HRC	DFA		FR: 0.08, 0.11, 0.14	NR: 0.8	
[69]	AISI H11	BBD	Ceramic	CS: 100, 130, 160	DOC: 0.1, 0.3, 0.5	RS: −530 to +625
	48–50 HRC			FR: 0.05, 0.125, 0.2	NR: 0.4, 0.8, 1.2	

(continued)

Table 2.2 (continued)

Authors	Cutting parameters					Performance characteristics
	Work material and hardness (HRC)	Methods (experiment and optimization)	Tool material	CS (m/min), FR (mm/rev)	DOC (mm), NR (mm)	CF (N), TW (mm), SR (μm), RS (MPa), MRR (mm^3/min)
[70]	Ni cast iron	Taguchi	Ceramic and CBN	CS: 50, 100, 150	DOC: 0.25, 0.5, 0.75	CF: 54–297 SR: 0.28–0.53
	62 HRC			FR: 0.05, 0.075, 0.1	NR: 0.08 (honed)	
[71]	AISI D3	FFD	Ceramic (Al$_2$O$_3$/TiC)	CS: 90, 120, 140	DOC: 0.15, 0.3, 0.45	CF: 65–371 SR: 0.37–3.45
	60 HRC	DFA		FR: 0.08, 0.16, 0.24	NR: 0.8	
[72]	AISI 420	CCD	Carbide (TiAlN)	CS: 100, 130, 170	DOC: 0.4	CF: 80–125 SR: <0.8
	48 HRC	DFA		FR: 0.1, 0.125, 0.16		
[73]	AISI D3	Taguchi	Carbide (TiSiN-TiAlN)	CS: 130, 155, 180	DOC: 0.1, 0.25, 0.4	TW: 0.098–0.2 SR: 0.51–1.15
	58 HRC	DFA		FR: 0.05, 0.1, 0.15	NR: 0.1–1.2	
[74]	AISI 52100	CCD	CC650 (Al$_2$O$_3$ + TiC)	CS: 100, 150, 200	DOC: 0.05, 0.15, 0.25	CF: 18.16–142 SR: 0.3–1.09
	59 HRC	DFA		FR: 0.08, 0.11, 0.14	NR: 0.8, 1.2, 1.6	
[75]	AISI 4140	FFD	Ceramic (Al$_2$O$_3$ + TiCN)	CS: 100, 170, 240	DOC: 0.1, 0.2, 0.3	SR: 0.71–1.93
	52 HRC	DFA		FR: 0.05, 0.1, 0.15	NR: 0.8	
[76]	Inconel 718	Taguchi	CBN	CS: 60, 80, 100, 120	DOC: 0.15, 0.3, 0.6	RS: 81–568.2
	47 HRC	GA		FR: 0.015–0.045		

(continued)

Table 2.2 (continued)

Authors	Cutting parameters					Performance characteristics
	Work material and hardness (HRC)	Methods (experiment and optimization)	Tool material	CS (m/min), FR (mm/rev)	DOC (mm), NR (mm)	CF (N), TW (mm), SR (μm), RS (MPa), MRR (mm^3/min)
[77]	AISI H11	Taguchi	CBN7020	CS: 120, 180	DOC: 0.15, 0.3, 0.45	SR: 0.18–0.565
	50 HRC	GRA		FR: 0.08–0.16		
[78]	AISI 52100	FFD	CBN	CS: 210, 260	DOC: 0.005, 0.01	SR: 0.1–0.2 RS: −120 to −400
	61 ± 1 HRC			FR: 0.05–0.1	NR: 0.8	
[79]	AISI 1060	FFD	Carbide	CS: 58, 81, 115	DOC: 1	CT: 700–1150
	40, 48, 56 HRC			FR: 0.1, 0.12, 0.14		
[80]	AISI 4140	Taguchi	CC6050 Ceramic	CS: 90, 120, 180	DOC: 0.15, 0.3, 0.45	SR: 0.23–0.63
	56 HRC	DFA		FR: 0.08, 0.12, 0.16	NR: 0.8	
[81]	AISI D2	Taguchi	PCBN	CS: 75–225	DOC: 0.1, 0.2, 0.3	CF: 66.8–149.61
	51, 58, 64 HRC			FR: 0.1, 0.2, 0.3	NR: 0.8–1.6	
[82]	AISI 1060	Taguchi	Carbide (TiCN, WC, Co)	CS: 81, 115, 161	NR: 0.8	CT: 835–1212 SR: 0.67–2.15
	40, 48, 56 HRC			FR: 0.12, 0.14, 0.16		
[83]	AISI H11	Taguchi	CC6050 CC650	CS: 100, 150, 200	DOC: 0.1, 0.3, 0.5	SR: 0.2–1.51
	50 HRC	DFA		FR: 0.08, 0.14, 0.2	NR: 0.8, 1.2	
[84]	AISI 4340	CCD	Ceramic	CS: 100, 160, 220	DOC: 0.1, 0.2, 0.3	SR: 0.35–1.03
	49 HRC	PSO		FR: 0.05, 0.10, 0.15	NR: 0.8, 1.2, 1.6	
[85]	AISI D3	Taguchi	CC650	CS: 200, 307, 440	DOC: 0.15, 0.3, 0.45	SR: 0.33–1.01 CF: 54–232

(continued)

Table 2.2 (continued)

Authors	Cutting parameters					Performance characteristics
	Work material and hardness (HRC)	Methods (experiment and optimization)	Tool material	CS (m/min), FR (mm/rev)	DOC (mm), NR (mm)	CF (N), TW (mm), SR (μm), RS (MPa), MRR (mm^3/min)
	60 HRC			FR: 0.08, 0.12, 0.16	NR: 0.8, 1.2, 1.6	
[38]	AISI H11	BBD	CBN7020	CS: 120, 180, 240	DOC: 0.15, 0.3, 0.45	CP: 3501–12228 P:173–996
	40, 45, 50 HRC	DFA		FR: 0.08, 0.12, 0.16		
[86]	AISI 1060	FFD	Carbide (TiCN, WC, Co)	CS: 58, 81, 115, 161	NR: 0.8	SR: 0.62
	40, 48, 56 HRC	SA		FR: 0.1, 0.12, 0.16, 0.18		
[87]	AISI 420	Taguchi	CC6050	CS: 80–340	DOC: 0.1–0.5	SR: 0.39–3.67 CF: 40.2–442
	59 HRC	DFA		FR: 0.08–0.4	NR: 0.8	
[39]	AISI H13	CCD	CBN	CS: 120, 150, 180	DOC: 0.08, 0.13, 0.18	SR: 0.14–1.89
	45, 50, 55 HRC	DFA		FR: 0.05, 0.1, 0.15	NR: 0.8	CF: 26–123
[88]	AISI 304	Taguchi	Carbide (TiCN + Al_2O_3)	CS: 150, 170, 190, 210	DOC: 0.5, 1, 1.5, 2	SR: <1.06; TW: 0.16; MRR: 29249
	48 HRC			FR: 0.15, 0.2, 0.25, 0.3	NR: 0.4, 0.8	
[89]	AISI 4340	Taguchi	Ceramic	CS: 110, 160, 210, 260	DOC: 0.2, 0.3, 0.4, 0.5	SR: 0.34–0.96
	49 HRC	DFA		FR: 0.06, 0.1, 0.18		
[90]	AISI 52100	CCD	Carbide	CS: 140, 195, 250	DOC: 0.08, 0.1, 0.11	TW: <0.3; SR: 0.8–2.1; CF: 15–83
	55 HRC	DFA		FR: 0.08, 0.1, 0.12		

Box–Behnken design (*BBD*); central composite design (*CCD*); cutting temperature (*CT*) in °C; cutting pressure (*CP*); desirability function approach (*DFA*); full factorial design (*FFD*); genetic algorithm (*GA*); grey relational analysis (*GRA*); material removal rate (*MRR*); power (*P*); residual stress (*RS*); simulated annealing (*SA*); zirconia toughened alumina (*ZTA*)

4. Taguchi method is suitable to optimize the single response, whereas alternate methods are utmost essential for optimizing the multiple responses simultaneously.
5. Taguchi method is not capable in establishing response equations (that is, input–output relations). These response equations are essential in making meaningful analysis of the process and optimization.

2.4.2 Response Surface Methodology (RSM)

RSM is a collection of statistical and mathematical tools that is aimed to solve complex multiple factor engineering problems with limited experimental runs, resources, efforts, cost, and time. RSM maps the input–output relationship of a process for the collected experimental data. Taguchi method limitations (such as failed to test all interaction factor effects, identifying nonlinear relationship, and deriving regression equations) can be overcome effectively by RSM [97, 98]. Central composite design (CCD) and Box–Behnken design (BBD) were used by many researchers as modelling and optimization tool to solve multi-factor problem of hard turning process [38, 39, 65, 69, 72, 74, 84, 90].

2.4.2.1 Central Composite Design (CCD)

CCD is an appropriate tool, which will fit the data points to a second-order polynomial function, and can be applied to optimize many research problems. CCD involves the combination of three groups, namely (a) two-level factorial or fractional factorial design (2^K) points which uses the set of -1 and $+1$ levels of a factor, (b) axial terms (star points) located axially at a distance (say α) from the centre to estimate quadratic terms, and (c) centre point experiments or replications that enable investigator to determine the experimental error.

CCD-based experiments are decided based on Eq. 2.1.

$$N = K^2 + 2K + n$$

Total Experiments $=$ Factors^2 $+ 2 *$ Factor $+$ number of replicates   (2.1)

In CCD, the value of alpha determines the location of axial points in experimental design and type of design such as spherical, orthogonal, rotatable, and face centred. Alpha value corresponds to that one dictates the position of axial point within the factorial portion region and leads to non-rotatable design. Face-centred central composite design with non-rotatable design was widely used in hard turning process [74, 84]. CCD matrices are available with rotatable (minimum five levels for a factor) and non-rotatable (minimum three levels for a factor) to develop full quadratic models (that is, response equations). The decision on selecting the type of design (i.e.

rotatable and non-rotatable) is made according to the geometric nature and practical constraints [99]. Thus, until some practical constraint dictates, the rotatable designs are not to be selected. The significant advantages of CCD over BBD are lowest variance at a distance from the centre and possibility to estimate the responses at the corners of a design space [100]. The optimal design points are confirmed better in estimating the response near the corners of a design matrix and worse near the centre of the design [100].

2.4.2.2 Box–Behnken Design (BBD)

BBD is a rotatable or nearly rotatable second-order response surface design, based on three-level fractional factorial designs. BBD offers significant advantage such that all factors are not varied simultaneously at their respective high or low values. This enables the design to avoid conducting experiments at their extreme condition. However, this design is concentrated more towards the centre and may not provide satisfactory results always. BBD requires the experiments to be conducted with minimum of five levels of variable. BBD-based experiments are decided based on Eq. 2.2,

$$N = 2K(k-1) + C$$

$$\text{Total Experiments} = 2\text{Factors (Factor} - 1) + \text{Centre Experiments} \qquad (2.2)$$

BBDs operate at their respective three levels considering multi-factors which are widely employed to model and optimize the hard turning process [38, 65, 69]. BBD requires only 13 experimental trials for the same three factors operating at respective three levels.

2.4.2.3 Full Factorial Design (FFD)

FFD considers all input variables to prepare the experimental or design matrices. The factorial points in the design matrix are treated as vertices of n-dimensional cube which are impending from FFD. FFD performs accurate analysis in determining the main and interaction factor effects. However, the time, effort, and cost involved in conducting experiments increase with the increase in number of factors. It is important to note that increased parameters and their respective levels tend to increase the requirements of practical experiments. Fractional factorial design (i.e. CCD, BBD, and Taguchi) is derived from a full factorial experimental matrix after utilizing the selected alias structure. The experimental size for three factors (X_1, X_2, and X_3) operating at their respective three levels (low, medium, and high) is found equal to 27 ($3^3 = \text{levels}^{\text{Factors}}$). The geometric representation of experimental points could appear on the vertex of a cube, corners or edges, origin, and middle of the faces [98].

2.4.3 Desirability Function Approach (DFA)

In 1980, Derringer and Suich developed desirability function approach method, used to optimize the multiple outputs. BBD, CCD, and Taguchi methods are used to develop in-process input–output relationship; however, they are not suitable to locate the extreme values of the response (that is, optimization). DFA overcomes these short-comings in performing the task of optimization. The response surfaces are estimated for each response separately. Here, the predicted values are determined for individual response surface and converted towards the dimensionless term d_i. The desirability value of the response surface varies in the range of zero (undesired response) and one (completely desirable). The composite desirability (D) value of the individual responses, required for multi-response optimization, is solved by using Eq. 2.3.

$$D = \left(d_1^{w_1} \times d_2^{w_2} \times \cdots d_m^{w_m} \right)^{1/m} \tag{2.3}$$

$$\text{Composite Desirability} = \left(\text{ind. desirability}_1^{\text{weight}_1} \times \text{ind.desirability}_2^{\text{weight}_2} \right.$$
$$\left. \times \cdots d_{m(\text{no. of responses})}^{1/m} \right)$$

In many cases, multi-objective optimization is applied to optimize the conflicting responses (i.e. both maximization and minimization). The solution corresponding to both the responses will require mathematical formulation. If the output function is of maximization type, then Eq. 2.4 is used.

$$Y_x = \frac{x - x_{\min}}{x_{\max} - x_{\min}} \tag{2.4}$$

Similarly, minimization type of the response function will follow Eq. 2.5.

$$Y_x = \frac{x_{\max} - x}{x_{\max} - x_{\min}} \tag{2.5}$$

Terms $x_{\max}$ and $x_{\min}$ are the maximum and the minimum values of the response χ. Term Y_x is the output value corresponding to desirability value of the response.

DFA is a simple and flexible tool, which works with intuitive concepts or judgements to solve both single and multi-objective optimization. DFA was applied to hard turning process by many researchers [38, 39, 66–68, 73–75, 80, 83, 87, 89, 90]. However, a few limitations of DFA method are listed below.

1. Trade-offs (i.e. weights) are the prerequisites to be assigned for the individual responses in DFA. DFA in many cases assign equal importance (i.e. weight fraction) for all responses. However, in real world practical problems assigning an equal importance to all responses may not be the practical requirement and may lead to misleading results.

2. DFA locates the solutions using deterministic search path with specific transition rules, and moving one solution with other could result in many local solutions.

2.4.4 Soft Computing Optimization Tools

Soft computing tools are stochastic in nature and work with the desired sets of probabilistic and transition rules. The optimal solutions are estimated in a multidimensional search space, thus enabling to target the global solutions always. Soft computing tools will conduct heuristic search and generate many potential or global solutions with reduced computation effort and cost. These techniques are derivative free and are not reliant on the in-process input–output relationships. Simulated annealing (SA) was used to formulate the optimization of hard turning process to attain minimum surface roughness by varying cutting speed, feed rate, and material hardness [86]. Particle swarm optimization (PSO) was attempted to minimize the surface roughness by considering cutting speed, axial feed, depth of cut, and nose radius [84]. The soft computing algorithms are found superior over iterative mathematical search techniques (i.e. linear programming, goal programming, and so on) and statistical optimization tools (DFA, GRA, and so on). This is due to their greater stability to solve convex multi-modal objective functions involving both continuous and discrete types of objective functions [2]. Note that the global solutions, in population-based search techniques, are reliant mainly on appropriate choice of algorithm-specific parameters (that is, crossover, mutation, population size and generations for GA, inertia weight, swarm size, iteration number for PSO, scout, employed and onlooker bees for ABC). It is interesting to note that teaching–learning-based optimization (TLBO) and Jaya algorithm do not require the tuning of algorithm-specific parameters to hit the global solutions [101, 102]. It is to be noted that the soft computing tools (i.e. GA, PSO, TLBO, Jaya) are treated as an appropriate tool to optimize the manufacturing processes [103–106]. Hence, there is a lot of scope to apply soft computing tools in modelling and optimization of hard turning process. A few limitations of soft computing-based optimization are as follows:

1. The convergence of soft computing tools may not guarantee to hit the global solution always.
2. No specific reference standards exist for the selection of optimization algorithm parameters (say inertia weight, swarm size, and generation for PSO).
3. Increased number of algorithm steps and parameters could not only increase the complexity but also the cost, effort, and time.

2.5 Capabilities of Hard Turning Process

A few observations, associated with hard turning process, are collected from the literature (refer Tables 2.1 and 2.2) and summarized in the following paragraphs:

1. Hard turning of extremely hard materials (i.e. materials with hardness more than 45 HRC) generates large amount of cutting forces and heat. Hence, the machine tool should possess high rigidity to limit the possible vibrations during machining. In addition, high rigidity in machine tool must enable to operate under high cutting speed without affecting the quality of machined surface finish.

2. Machining hard materials will generate large cutting forces and heat; thereby, the cutting tool must withstand the mechanical wear (adhesion and abrasion) and thermal shock. Cutting tools such as CBN, PCBN, ceramic, carbides, and coated tool materials are generally used due to their excellent machinability characteristics. It was observed that the application of coating on the cutting tool reduces the cutting forces and temperatures. This resulted in improved surface finish and tool life.

3. Hard turning process was carried out mostly under dry cutting environment. This will limit the cost of cutting fluid and its associated maintenance to around 17–30%. However, dry cutting environment will generate high temperature as a result of friction at tool–work material interface and result in adhesion and rapid tool wear. To limit this, the tool material should retain hardness (i.e. hot hardness) at high temperature and is chemically immune (i.e. diffusion). In addition, the formation of white layers will cause thermal distortion on the machined surface (i.e. surface integrity) and dimensional inaccuracy can be reduced with the application of coolants. The coolant can reduce the surface roughness on machined part and improve the tool life by reducing wear.

4. Increase of work material hardness from 45 to 50 HRC gradually increases cutting forces. However, the work material hardness above 50 HRC showed a rapid increase in cutting forces and results in segmented chips.

2.5.1 Variables of Hard Turning Process

Various modelling and optimization tools (OFAT, RSM, Taguchi, numerical and soft computing tools) applied to machining of hard materials (above 45 HRC) have shown that machining parameters (cutting speed, feed rate, nose radius, depth of cut) have direct influence on machining performances. It is to be noted that researchers have selected different ranges of machining parameters for the materials with same hardness value [60, 68, 69, 76, 85, 89]. It is also observed that till date there are no universal standards defined to determine the range of process variables. Further, the operating range of machining parameters is also dependent on the tool material used in machining. Note that the selected machining parameters with narrow range (levels) could result in poor process information. On the other hand, a wide range of operating levels may lead to infeasible solutions. The primary motto of machining engineer is to determine the parameters and their operating values that could maximize productivity and surface integrity, while minimizing the cost and energy consumption. Attempts are made to provide detailed insights of summary of machining parameters along with their operating ranges for the material with hardness 45–69 HRC. The process variables, which influence the machining performances, are discussed below.

2.5.1.1   Cutting Speed

The operating level of the cutting speed for turning the hard materials with a different hardness is presented in Fig. 2.2. The highest cutting speed of 600 m/min is observed for the material, possessing hardness of 47–48 HRC. It is to be noted that the cutting speed is reduced with the increase in work material hardness to keep the tool wear minimum. Coated carbide and ceramic tool materials are used for the cutting speed less than 200 m/min. CBN tools are generally used for machining hard materials (that is, above 60 HRC) with a cutting speed above 200 m/min. It is also observed that the operating range of cutting speed is maintained within 150 m/min for the materials with hardness value above 65 HRC, irrespective of the type of cutting tool material used. Cutting forces will tend to decrease at high temperature associated with increase in cutting speed [107]. This is due to softening of work material at high temperature. Furthermore, the built-up edge formation is eliminated which could improve the surface finish when the machine is operating at higher cutting speed. Although higher cutting speed results in a better surface finish, it affects tool wear [52]. At low cutting

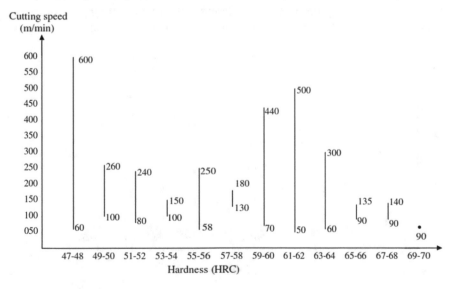

Fig. 2.2 Cutting speed versus material hardness

speed, the cutting forces are higher which could result in large amount of built-up edge formation. It consequently induces tool wear, deteriorates the surface finish, and increases the machining time and cost [108]. The better surface finish with high cutting speed is attributed to gradual reduction in grooves that develop on flank face or wear land [109]. In addition, cutting speed (approximately 78%) contributions are more towards power consumption. Power consumption will be increased drastically when cutting speed is increased along with feed and depth of cut [64]. Keeping in view of tool life, machined surface quality, machining time, power consumption, and cost, determining the optimum cutting speed is always desirable.

2.5.1.2 Feed Rate

Figure 2.3 shows the operating levels of feed rate employed in hard turning process to machine materials with a different hardness. Generally, the feed rate varies in the range of 0.015–0.4 mm/rev, depending upon the hardness of material. Many researchers used a feed rate of 0.3 mm/rev for the materials with hardness up to 65 HRC. However, the feed rate is restricted to less than 0.16 mm/rev for materials with hardness more than 65 HRC. It was observed that the ceramic tools are not advisable for the feed rate above 0.25 mm/rev. This might be due to its high brittleness property. Furthermore, many research efforts have shown that the carbon boron nitride and coated carbide tools are used as tool materials for high feed rate. As the feed rate decreases, the surface finish on the machined part improves, but it possesses negative impact on tool life [52]. Note that the increased values of feed could yield high values of compressive stresses, high power consumption, reduced material removal and cost [109, 110]. At larger feed rate, the workpiece will offer greater resistance to the tool in the cutting direction, which increases the friction, cutting forces, power consumption, and tool wear [111]. However, high feed rate is prerequisite to lower the

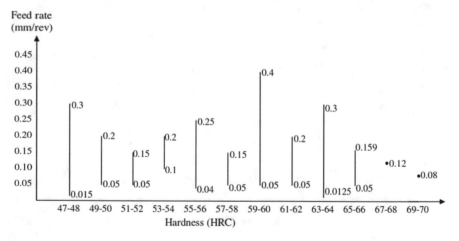

Fig. 2.3 Feed rate versus material hardness

specific cutting force [111]. Note that beyond the critical limit of feed, the machining cost increases due to increased tool cost. In addition, the machined surface quality deteriorates, which requires additional pass with low feed or secondary processes (electropolishing, grinding, and so on). There exists a direct relationship between the machined surface characteristics and the functional performances such as fatigue strength, corrosion, and other tribology properties [109]. Monitoring the hard turning process requires optimum value of feed rate. Use of optimal feed rate will improve tool life, reduce machining cost, reduce energy consumption, and improve surface integrity.

2.5.1.3 Depth of Cut

Figure 2.4 shows the levels of cutting parameter and depth of cut used to machine materials having different hardness. It has been observed that the depth of cut is observed to vary in the ranges of 0.005–0.75 mm (except 2 mm in one literature). In general, high values of depth of cut are found impractical as they not only reduce the tool life, but also deteriorate the surface roughness. If there is an increase in depth of cut, the cutting velocity will increase (due to increased cutting-edge angle), which in turn increases forces and energy consumption (i.e. power consumption) in machining of hard materials [64]. Experimental studies have shown that the depth of cut contributions is more for cutting forces as compared to feed rate and cutting speed [107]. With the increased values in depth of cut, the surface finish and tool life will deteriorate due to chattering of cutting tool [109]. In general, higher depth of cut is desirable in view of machining time as a result of high material removal. CBN tools are widely used when higher depth of cut is required (refer Tables 1 and 2). The researchers are more interested to know the best value of depth of cut with a focus on material hardness, tool material and wear, tool life, surface integrity, and machining cost.

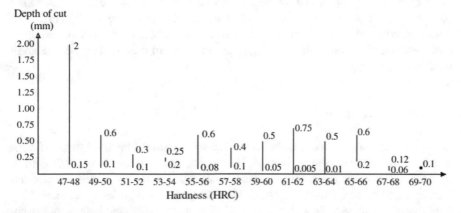

Fig. 2.4 Depth of cut versus material hardness

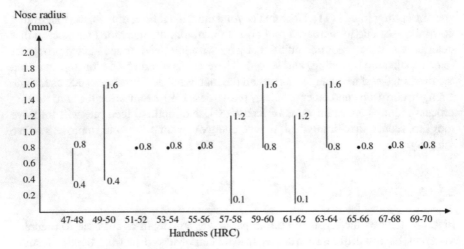

Fig. 2.5 Tool nose radius versus material hardness

2.5.1.4   Nose Radius

The cutting forces, tool wear, and surface finish are significantly influenced by the nose radius. Figure 2.5 shows the nose radius values and levels employed while machining materials having a different hardness (HRC). Note that the values of nose radius generally used in conducting experiments are varied in the ranges of 0.1–1.6 mm. Experimental results have shown that larger tool nose radii will tend to improve surface finish, whereas the tool wear and specific cutting energy are higher as compared to small nose radii [112]. Note that many research efforts used the constant nose radius of 0.8 mm (refer Fig. 2.5). Higher nose radii will enable the industry personnel to operate with higher feed rate and minimize tool wear without affecting the surface roughness [113]. Large nose radius affects the cutting forces, alters chip morphology, and minimizes the uncut chip thickness. Further, it increases ploughing effect in cutting zone, which leads to high flank wear [49, 114]. Low values of nose radii tend to make rough surface on the machined part and increase tool wear quickly. Hence, determining optimum value of nose radius is desirable in hard turning process.

2.5.1.5   Tool Geometry

Tool geometry having chamfered cutting edge will produce improved surface finish, provided the material should have high hardness and operate under high cutting speed. On the other hand, inverse trend is observed with low values of material hardness and cutting speed [52]. Negative rake angles are preferred in machining hard materials, since tools with negative rake angle can withstand high compressive forces [110]. Experimental studies have shown that the honed edge geometry will result in better surface finish and reduced tool wear [115]. Hone radius with varied edge will reduce the heat generation along the cutting-edge direction and tend to induce less plastic strain on the workpiece and reduce tool wear [116]. Tool life of the chamfered cutting edge is not appreciable, whereas better tool life is expected from the honed radius [108]. However, comparatively higher friction at secondary cutting edge is experienced by hone radii due to the low chip thickness-to-edge radius ratio. In view of the above, hone radii coupled with negative rake angle is better for hard turning process.

2.6 Conclusion

Many research efforts are made to improve performance characteristics in hard turning process. In most of the studies, the researchers have attempted to optimize the influencing parameters such as tool geometry, materials and coatings, material hardness, cutting parameters (cutting speed, feed rate, nose radius, depth of cut, and so on) for better machining performance parameters (i.e. tool wear, surface roughness, material removal rate, cutting forces, residual stresses, and so on). To attain better machinability index, various approaches, namely analytical/numerical, one factor at a time, statistical design of experiments (FFD, CCD, BBD, Taguchi), are widely used. A combination of response surface methodology and statistical analysis is a powerful tool that can be applied to manufacturing processes to analyse the process completely and make considerable improvements.

This chapter concludes with the review of a different methodology employed in hard turning. Further, it will also help the industry personnel to know the appropriate tool and work material combinations and their properties. In addition, the review carried out will also help the researchers to select the optimal range of the cutting parameters based on material hardness in modelling and optimization of machining process. It is to be noted that only few research efforts are published which specifically focussed on predicting of machining parameters. Deriving empirical equations that are based on experimental data is of industrial relevance and will also assist the researchers working in the area of machining. Therefore, comprehensive analysis, modelling, prediction, and optimization of parameters are needed for further studies in hard turning process.

References

1. P. Ettmayer, H. Kolaska, H.M. Ortner, History of hardmetals. Compr. Hard Mater. **1**, 3–27 (2014). https://doi.org/10.1016/B978-0-08-096527-7.00001-5
2. I. Mukherjee, P.K. Ray, A review of optimization techniques in metal cutting processes. Comput. Ind. Eng. **50**(1–2), 15–34 (2006)
3. Y. Huang, S.Y. Liang, Cutting forces modeling considering the effect of tool thermal property—application to CBN hard turning. Int. J. Mach. Tools Manuf. **43**(3), 307–315 (2003)

4. Y. Huang, S.Y. Liang, Modeling of cutting forces under hard turning conditions considering tool wear effect. J. Manuf. Sci. Eng. **127**(2), 262–270 (2005)
5. P.J. Arrazola, T. Ozel, Numerical modelling of 3D hard turning using arbitrary Lagrangian Eulerian finite element method. Int. J. Mach. Mach. Mater. **4**(1), 14–25 (2008)
6. C. Scheffer, H. Kratz, P.S. Heyns, F. Klocke, Development of a tool wear-monitoring system for hard turning. Int. J. Mach. Tools Manuf **43**(10), 973–985 (2003)
7. J.S. Dureja, V.K. Gupta, V.S. Sharma, M. Dogra, M.S. Bhatti, A review of empirical modeling techniques to optimize machining parameters for hard turning applications. Proc. Inst. Mech. Eng. Part B: J. Eng. Manuf. **230**(3), 389–404 (2016)
8. F.J. Pontes, A.P. de Paiva, P.P. Balestrassi, J.R. Ferreira, M.B. da Silva, Optimization of radial basis function neural network employed for prediction of surface roughness in hard turning process using Taguchi's orthogonal arrays. Expert Syst. Appl. **39**(9), 7776–7787 (2012)
9. A. Agrawal, S. Goel, W.B. Rashid, M. Price, Prediction of surface roughness during hard turning of AISI 4340 steel (69 HRC). Appl. Soft Comput. **30**, 279–286 (2015)
10. M. Sayuti, A.A. Sarhan, F. Salem, Novel uses of SiO_2 nano-lubrication system in hard turning process of hardened steel AISI4140 for less tool wear, surface roughness and oil consumption. J. Clean. Prod. **67**, 265–276 (2014)
11. K. Bouacha, M.A. Yallese, S. Khamel, S. Belhadi, Analysis and optimization of hard turning operation using cubic boron nitride tool. Int. J. Refract Metal Hard Mater. **45**, 160–178 (2014)
12. F. Klocke, E. Brinksmeier, K. Weinert, Capability profile of hard cutting and grinding processes. CIRP Ann. Manuf. Technol. **54**(2), 22–45 (2005)
13. B.P. Erdel, *High-Speed Machining* (Society of Manufacturing Engineers, 2003)
14. R. Suresh, S. Basavarajappa, V.N. Gaitonde, G.L. Samuel, J.P. Davim, State-of-the-art research in machinability of hardened steels. Proc. Inst. Mech. Eng. Part B: J. Eng. Manuf. **227**(2), 191–209 (2013)
15. V.P. Astakhov, Machining of hard materials–definitions and industrial applications, *Machining of Hard Materials* (Springer, London, 2011), pp. 1–32
16. C. Daniel, One-at-a-time plans. J. Am. Stat. Assoc. **68**(342), 353–360 (1973)
17. M. Friedman, L.J. Savage, Planning experiments seeking maxima. Tech. Stat. Anal. (1947), pp. 365–372
18. Z. Li, B. Chen, H. Wu, X. Ye, B. Zhang, A design of experiment aided stochastic parameterization method for modeling aquifer NAPL contamination. Environ. Model Softw. **101**, 183–193 (2018)
19. G.E. Box, P.Y. Liu, Statistics as a catalyst to learning by scientific method part I—an example. J. Qual. Technol. **31**(1), 1–15 (1999)
20. V. Czitrom, One-factor-at-a-time versus designed experiments. Am. Stat. **53**(2), 126–131 (1999)
21. C.J. Wu, M.S. Hamada, *Experiments: Planning, Analysis, and Optimization*, vol. 552 (Wiley & Sons, 2011)
22. D.D. Frey, H. Wang, Adaptive one-factor-at-a-time experimentation and expected value of improvement. Technometrics **48**(3), 418–431 (2006)
23. G.C. Manjunath Patel, P. Krishna, M.B. Parappagoudar, Modelling and multi-objective optimisation of squeeze casting process using regression analysis and genetic algorithm. Aust. J. Mech. Eng. **14**(3), 182–198 (2016)
24. G.C. Manjunath Patel, P. Krishna, M.B. Parappagoudar, Modelling in squeeze casting process-present state and future perspectives. Adv. Autom. Eng. **4**(1), 1–9 (2015)
25. R.T. Coelho, E.G. Ng, M.A. Elbestawi, Tool wear when turning hardened AISI 4340 with coated PCBN tools using finishing cutting conditions. Int. J. Mach. Tools Manuf. **47**(2), 263–272 (2007)
26. J.L. Li, L.L. Jing, M. Chen, An FEM study on residual stresses induced by high-speed end-milling of hardened steel SKD11. J. Mater. Process. Technol. **209**(9), 4515–4520 (2009)
27. J. Lorentzon, N. Järvstråt, Modelling tool wear in cemented-carbide machining alloy 718. Int. J. Mach. Tools Manuf. **48**(10), 1072–1080 (2008)

28. H.J. Hu, W.J. Huang, Studies on wears of ultrafine-grained ceramic tool and common ceramic tool during hard turning using Archard wear model. Int. J. Adv. Manuf. Technol. **69**(1–4), 31–39 (2013)
29. D.M. Kim, V. Bajpai, B.H. Kim, H.W. Park, Finite element modeling of hard turning process via a micro-textured tool. Int. J. Adv. Manuf. Technol. **78**(9–12), 1393–1405 (2015)
30. C.S. Kumar, S.K. Patel, Application of surface modification techniques during hard turning: present work and future prospects. Int. J. Refract Metal Hard Mater. **76**, 112–127 (2018)
31. C.S. Kumar, S.K. Patel, Effect of chip sliding velocity and temperature on the wear behaviour of PVD AlCrN and AlTiN coated mixed alumina cutting tools during turning of hardened steel. Surf. Coat. Technol. **334**, 509–525 (2018)
32. L. Ma, C. Li, J. Chen, W. Li, Y. Tan, C. Wang, Y. Zhou, Prediction model and simulation of cutting force in turning hard-brittle materials. Int. J. Adv. Manuf. Technol. **91**(1–4), 165–174 (2017)
33. C. Shet, X. Deng, Residual stresses and strains in orthogonal metal cutting. Int. J. Mach. Tools Manuf. **43**(6), 573–587 (2003)
34. K. Li, X.L. Gao, J.W. Sutherland, Finite element simulation of the orthogonal metal cutting process for qualitative understanding of the effects of crater wear on the chip formation process. J. Mater. Process. Technol. **127**(3), 309–324 (2002)
35. F. Akbar, P.T. Mativenga, M.A. Sheikh, An experimental and coupled thermo-mechanical finite element study of heat partition effects in machining. Int. J. Adv. Manuf. Technol. **46**(5–8), 491–507 (2010)
36. A. Qasim, S. Nisar, A. Shah, M.S. Khalid, M.A. Sheikh, Optimization of process parameters for machining of AISI-1045 steel using Taguchi design and ANOVA. Simul. Model. Pract. Theory **59**, 36–51 (2015)
37. M.E. Korkmaz, M. Günay, Finite element modelling of cutting forces and power consumption in turning of AISI 420 martensitic stainless steel. Arab. J. Sci. Eng. (2018), pp. 1–8
38. S. Benlahmidi, H. Aouici, F. Boutaghane, A. Khellaf, B. Fnides, M.A. Yallese, Design optimization of cutting parameters when turning hardened AISI H11 steel (50 HRC) with CBN7020 tools. Int. J. Adv. Manuf. Technol. **89**(1–4), 803–820 (2017)
39. P. Kumar, S. Chauhan, C. Pruncu, M. Gupta, D. Pimenov, M. Mia, H. Gill, Influence of different grades of CBN inserts on cutting force and surface roughness of AISI H13 die tool steel during hard turning operation. Materials **12**(1), 177 (2019)
40. G.E. Box, N.R. Draper, *Empirical Model-Building and Response Surfaces* (Wiley & Sons, 1987)
41. E. Budak, Y. Altintas, E.J.A. Armarego, Prediction of milling force coefficients from orthogonal cutting data. J. Manuf. Sci. Eng. **118**(2), 216–224 (1996)
42. A.S. More, W. Jiang, W.D. Brown, A.P. Malshe, Tool wear and machining performance of cBN–TiN coated carbide inserts and PCBN compact inserts in turning AISI 4340 hardened steel. J. Mater. Process. Technol. **180**(1–3), 253–262 (2006)
43. J.A. Arsecularatne, L.C. Zhang, C. Montross, P. Mathew, On machining of hardened AISI D2 steel with PCBN tools. J. Mater. Process. Technol. **171**(2), 244–252 (2006)
44. Y.K. Chou, C.J. Evans, M.M. Barash, Experimental investigation on CBN turning of hardened AISI 52100 steel. J. Mater. Process. Technol. **124**(3), 274–283 (2002)
45. T.G. Dawson, T.R. Kurfess, Machining hardened steel with ceramic-coated and uncoated CBN cutting tools. Soc. Manuf. Eng. **156**, 1–7 (2002)
46. V.G. Navas, I. Ferreres, J.A. Marañón, C. Garcia-Rosales, J.G. Sevillano, Electro-discharge machining (EDM) versus hard turning and grinding—Comparison of residual stresses and surface integrity generated in AISI O1 tool steel. J. Mater. Process. Technol. **195**(1–3), 186–194 (2008)
47. M.A. Kamely, M.Y. Noordin, V.C. Venkatesh, The effect of multiple pass cutting on surface integrity when hard turning of AISI D2 cold work tool steel. Int. J. Precis. Technol. **1**(1), 97–105 (2007)
48. H.A. Kishawy, M.A. Elbestawi, Tool wear and surface integrity during high-speed turning of hardened steel with polycrystalline cubic boron nitride tools. Proc. Inst. Mech. Eng. Part B: J. Eng. Manuf. **215**(6), 755–767 (2001)

49. M. Liu, J.I. Takagi, A. Tsukuda, Effect of tool nose radius and tool wear on residual stress distribution in hard turning of bearing steel. J. Mater. Process. Technol. **150**(3), 234–241 (2004)
50. S.K. Khrais, Y.J. Lin, Wear mechanisms and tool performance of TiAlN PVD coated inserts during machining of AISI 4140 steel. Wear **262**(1–2), 64–69 (2007)
51. C.K. Toh, Tool life and tool wear during high-speed rough milling using alternative cutter path strategies. Proc. Inst. Mech. Eng. Part B: J. Eng. Manuf. **217**(9), 1295–1304 (2003)
52. T. Ozel, T.K. Hsu, E. Zeren, Effects of cutting edge geometry, workpiece hardness, feed rate and cutting speed on surface roughness and forces in finish turning of hardened AISI H13 steel. Int. J. Adv. Manuf. Technol. **25**(3–4), 262–269 (2005)
53. J.M. Zhou, H. Walter, M. Andersson, J.E. Stahl, Effect of chamfer angle on wear of PCBN cutting tool. Int. J. Mach. Tools Manuf. **43**(3), 301–305 (2003)
54. R. Meyer, J. Köhler, B. Denkena, Influence of the tool corner radius on the tool wear and process forces during hard turning. Int. J. Adv. Manuf. Technol. **58**(9–12), 933–940 (2012)
55. J. Bhaskaran, M. Murugan, N. Balashanmugam, M. Chellamalai, Monitoring of hard turning using acoustic emission signal. J. Mech. Sci. Technol. **26**(2), 609–615 (2012)
56. K. Aslantas, I. Ucun, A. Cicek, Tool life and wear mechanism of coated and uncoated Al2O3/TiCN mixed ceramic tools in turning hardened alloy steel. Wear **274**, 442–451 (2012)
57. W.B. Rashid, S. Goel, X. Luo, J.M. Ritchie, An experimental investigation for the improvement of attainable surface roughness during hard turning process. Proc. Inst. Mech. Eng. Part B: J. Eng. Manuf. **227**(2), 338–342 (2013)
58. A.K. Sahoo, B. Sahoo, Experimental investigations on machinability aspects in finish hard turning of AISI 4340 steel using uncoated and multilayer coated carbide inserts. Measurement **45**(8), 2153–2165 (2012)
59. R. Ferreira, J. Řehoř, C.H. Lauro, D. Carou, J.P. Davim, Analysis of the hard turning of AISI H13 steel with ceramic tools based on tool geometry: surface roughness, tool wear and their relation. J. Braz. Soc. Mech. Sci. Eng. **38**(8), 2413–2420 (2016)
60. G. Zheng, R. Xu, X. Cheng, G. Zhao, L. Li, J. Zhao, Effect of cutting parameters on wear behavior of coated tool and surface roughness in high-speed turning of 300M. Measurement **125**, 99–108 (2018)
61. I. Lazoglu, K. Buyukhatipoglu, H. Kratz, F. Klocke, Forces and temperatures in hard turning. Mach. Sci. Technol. **10**(2), 157–179 (2006)
62. A. Srithar, K. Palanikumar, B. Durgaprasad, Experimental investigation and surface roughness analysis on hard turning of AISI D2 steel using coated carbide insert. Procedia Eng. **97**, 72–77 (2014)
63. F. Puh, T. Šegota, Z. Jurković, Optimization of hard turning process parameters with PCBN tool based on the Taguchi method. Tehnički vjesnik **19**(2), 415–419 (2012)
64. R. Suresh, S. Basavarajappa, G.L. Samuel, Some studies on hard turning of AISI 4340 steel using multilayer coated carbide tool. Measurement **45**(7), 1872–1884 (2012)
65. H. Aouici, M.A. Yallese, K. Chaoui, T. Mabrouki, J.F. Rigal, Analysis of surface roughness and cutting force components in hard turning with CBN tool: Prediction model and cutting conditions optimization. Measurement **45**(3), 344–353 (2012)
66. H. Aouici, M.A. Yallese, A. Belbah, M.F. Ameur, M. Elbah, Experimental investigation of cutting parameters influence on surface roughness and cutting forces in hard turning of X38CrMoV5-1 with CBN tool. Sadhana **38**(3), 429–445 (2013)
67. Z. Hessainia, A. Belbah, M.A. Yallese, T. Mabrouki, J.F. Rigal, On the prediction of surface roughness in the hard turning based on cutting parameters and tool vibrations. Measurement **46**(5), 1671–1681 (2013)
68. M. Elbah, M.A. Yallese, H. Aouici, T. Mabrouki, J.F. Rigal, Comparative assessment of wiper and conventional ceramic tools on surface roughness in hard turning AISI 4140 steel. Measurement **46**(9), 3041–3056 (2013)
69. S. Saini, I.S. Ahuja, V.S. Sharma, Modelling the effects of cutting parameters on residual stresses in hard turning of AISI H11 tool steel. Int. J. Adv. Manuf. Technol. **65**(5–8), 667–678 (2013)

70. E. Yucel, M. Gunay, Modelling and optimization of the cutting conditions in hard turning of high-alloy white cast iron (Ni-Hard). Proc. Inst. Mech. Eng. Part C: J. Mech. Eng. Sci. **227**(10), 2280–2290 (2013)
71. H. Aouici, H. Bouchelaghem, M.A. Yallese, M. Elbah, B. Fnides, Machinability investigation in hard turning of AISI D3 cold work steel with ceramic tool using response surface methodology. Int. J. Adv. Manuf. Technol. **73**(9–12), 1775–1788 (2014)
72. M.Y. Noordin, D. Kurniawan, Y.C. Tang, K. Muniswaran, Feasibility of mild hard turning of stainless steel using coated carbide tool. Int. J. Adv. Manuf. Technol. **60**(9–12), 853–863 (2012)
73. J.S. Dureja, R. Singh, M.S. Bhatti, Optimizing flank wear and surface roughness during hard turning of AISI D3 steel by Taguchi and RSM methods. Prod. Manuf. Res. **2**(1), 767–783 (2014)
74. I. Meddour, M.A. Yallese, R. Khattabi, M. Elbah, L. Boulanouar, Investigation and modeling of cutting forces and surface roughness when hard turning of AISI 52100 steel with mixed ceramic tool: cutting conditions optimization. Int. J. Adv. Manuf. Technol. **77**(5–8), 1387–1399 (2015)
75. S.R. Das, D. Dhupal, A. Kumar, Study of surface roughness and flank wear in hard turning of AISI 4140 steel with coated ceramic inserts. J. Mech. Sci. Technol. **29**(10), 4329–4340 (2015)
76. F. Jafarian, H. Amirabadi, J. Sadri, Experimental measurement and optimization of tensile residual stress in turning process of Inconel718 superalloy. Measurement **63**, 1–10 (2015)
77. H. Aouici, B. Fnides, M. Elbah, S. Benlahmidi, H. Bensouilah, M. Yallese, Surface roughness evaluation of various cutting materials in hard turning of AISI H11. Int. J. Ind. Eng. Comput. **7**(2), 339–352 (2016)
78. P. Revel, N. Jouini, G. Thoquenne, F. Lefebvre, High precision hard turning of AISI 52100 bearing steel. Prec. Eng. **43**, 24–33 (2016)
79. M. Mia, N.R. Dhar, Response surface and neural network based predictive models of cutting temperature in hard turning. J. Adv. Res. **7**(6), 1035–1044 (2016)
80. H. Zahia, Y. Athmane, B. Lakhdar, M. Tarek, On the application of response surface methodology for predicting and optimizing surface roughness and cutting forces in hard turning by PVD coated insert. Int. J. Ind. Eng. Comput. **6**(2), 267–284 (2015)
81. L. Tang, Z. Cheng, J. Huang, C. Gao, W. Chang, Empirical models for cutting forces in finish dry hard turning of hardened tool steel at different hardness levels. Int. J. Adv. Manuf. Technol. **76**(1–4), 691–703 (2015)
82. M. Mia, N.R. Dhar, Optimization of surface roughness and cutting temperature in high-pressure coolant-assisted hard turning using Taguchi method. Int. J. Adv. Manuf. Technol. **88**(1–4), 739–753 (2017)
83. A. Khellaf, H. Aouici, S. Smaiah, S. Boutabba, M.A. Yallese, M. Elbah, Comparative assessment of two ceramic cutting tools on surface roughness in hard turning of AISI H11 steel: including 2D and 3D surface topography. Int. J. Adv. Manuf. Technol. **89**(1–4), 333–354 (2017)
84. A. Panda, S.R. Das, D. Dhupal, Surface roughness analysis for economical feasibility study of coated ceramic tool in hard turning operation. Proc. Integr. Optimization Sustain. **1**(4), 237–249 (2017)
85. O. Zerti, M.A. Yallese, R. Khettabi, K. Chaoui, T. Mabrouki, Design optimization for minimum technological parameters when dry turning of AISI D3 steel using Taguchi method. Int. J. Adv. Manuf. Technol. **89**(5–8), 1915–1934 (2017)
86. M. Mia, N.R. Dhar, Modeling of surface roughness using RSM, FL and SA in dry hard turning. Arab. J. Sci. Eng. **43**(3), 1125–1136 (2018)
87. A. Zerti, M.A. Yallese, O. Zerti, M. Nouioua, R. Khettabi, Prediction of machining performance using RSM and ANN models in hard turning of martensitic stainless steel AISI 420. Proc. Inst. Mech. Eng. Part C: J. Mech. Eng. Sci. (2019). https://doi.org/10.1177/0954406218820557

88. M. Kaladhar, Evaluation of hard coating materials performance on machinability issues and material removal rate during turning operations. Measurement **135**, 493–502 (2019)

89. J. Jena, A. Panda, A.K. Behera, P.C. Jena, S.R. Das, D. Dhupal, Modeling and optimization of surface roughness in hard turning of AISI 4340 steel with coated ceramic tool, in *Innovation in Materials Science and Engineering* (Springer, Singapore, 2019), pp. 151–160

90. A. Alok, M. Das, Multi-objective optimization of cutting parameters during sustainable dry hard turning of AISI 52100 steel with newly develop HSN2-coated carbide insert. Measurement **133**, 288–302 (2019)

91. M.S. Phadke, *Quality Enginuring using Robust Design* (Prentice Hall, New Jersey, 1989)

92. G. Taguchi, Y. Wu, Introduction to off-line quality control, Central Japan quality control association. Avail. Am. Suppl. Inst., vol. 32100 (1980)

93. K.L. Tsui, An overview of Taguchi method and newly developed statistical methods for robust design. Iie Transactions **24**(5), 44–57 (1992)

94. P.J. Ross, P.J. Ross, *Taguchi Techniques for Quality Engineering: Loss Function, Orthogonal Experiments, Parameter And Tolerance Design (No. TS156 R12)* (McGraw-Hill, New York, 1988)

95. R. Unal, E.B. Dean, Taguchi Approach To Design Optimization for Quality and Cost: An Overview

96. W.M. Carlyle, D.C. Montgomery, G.C. Runger, Optimization problems and methods in quality control and improvement. J. Qual. Technol. **32**(1), 1–17 (2000)

97. G.E. Box, K.B. Wilson, On the experimental attainment of optimum conditions. J. Royal Stat. Soc. Ser. B (Methodol.) **13**(1), 1–38 (1951)

98. D.C. Montgomery, *Design and Analysis of Experiments*, vol. 52 (Wiley & Sons, 2001), pp. 218–286

99. G.C.M. Patel, P. Krishna, M.B. Parappagoudar, Squeeze casting process modeling by a conventional statistical regression analysis approach. Appl. Math. Modell. **40**(15–16), 6869–6888 (2016)

100. L.A. Trinca, S.G. Gilmour, Difference variance dispersion graphs for comparing response surface designs with applications in food technology. J. Royal Stat. Soc. Ser. C (Appl. Stat.) **48**(4), 441–455 (1999)

101. R.V. Rao, V.J. Savsani, D.P. Vakharia, Teaching–learning-based optimization: a novel method for constrained mechanical design optimization problems. Comput. Aided Des. **43**(3), 303–315 (2011)

102. R.V. Rao, Single-and multi-objective optimization of casting processes using Jaya algorithm and its variants, in *Jaya: An Advanced Optimization Algorithm and its Engineering Applications* (Springer, Cham, 2019), pp. 273–289

103. G.R. Chate, G.C.M. Patel, A.S. Deshpande, M.B. Parappagoudar, Modeling and optimization of furan molding sand system using design of experiments and particle swarm optimization. Proc. Inst. Mech. Eng. Part E: J. Process Mech. Eng. **232**(5), 579–598 (2018)

104. G.R. Chate, G.C.M. Patel, S.B. Bhushan, M.B. Parappagoudar, A.S. Deshpande, Comprehensive modelling, analysis and optimization of furan resin-based moulding sand system with sawdust as an additive. J. Braz. Soc. Mech. Sci. Eng. **41**(4), 183 (2019)

105. G.C.M. Patel, P. Krishna, P.R. Vundavilli, M.B. Parappagoudar, Multi-objective optimization of squeeze casting process using genetic algorithm and particle swarm optimization. Arch. Foundry Eng. **16**(3), 172–186 (2016)

106. G.C.M. Patel, P. Krishna, M.B. Parappagoudar, P.R. Vundavilli, Multi-objective optimization of squeeze casting process using evolutionary algorithms. Int. J. Swarm Intell. Res. (IJSIR) **7**(1), 55–74 (2016)

107. K. Bouacha, M.A. Yallese, T. Mabrouki, J.F. Rigal, Statistical analysis of surface roughness and cutting forces using response surface methodology in hard turning of AISI 52100 bearing steel with CBN tool. Int. J. Refract. Metals Hard Mater. **28**(3), 349–361 (2010)

108. G. Bartarya, S.K. Choudhury, Effect of cutting parameters on cutting force and surface roughness during finish hard turning AISI52100 grade steel. Procedia CIRP **1**, 651–656 (2012)

109. V.S. Sharma, S. Dhiman, R. Sehgal, S.K. Sharma, Estimation of cutting forces and surface roughness for hard turning using neural networks. J. Intell. Manuf. **19**(4), 473–483 (2008)

110. P. Dahlman, F. Gunnberg, M. Jacobson, The influence of rake angle, cutting feed and cutting depth on residual stresses in hard turning. J. Mater. Process. Technol. **147**(2), 181–184 (2004)

111. V.N. Gaitonde, S.R. Karnik, L. Figueira, J.P. Davim, Analysis of machinability during hard turning of cold work tool steel (type: AISI D2). Mater. Manuf. Process. **24**(12), 1373–1382 (2009)

112. J. Hua, R. Shivpuri, X. Cheng, V. Bedekar, Y. Matsumoto, F. Hashimoto, T.R. Watkins, Effect of feed rate, workpiece hardness and cutting edge on subsurface residual stress in the hard turning of bearing steel using chamfer + hone cutting edge geometry. Mater. Sci. Eng. A **394**(1–2), 238–248 (2005)

113. A. Madariaga, J.A. Esnaola, E. Fernandez, P.J. Arrazola, A. Garay, F. Morel, Analysis of residual stress and work-hardened profiles on Inconel 718 when face turning with large-nose radius tools. Int. J. Adv. Manuf. Technol. **71**(9–12), 1587–1598 (2014)

114. M. Dogra, V.S. Sharma, J. Dureja, Effect of tool geometry variation on finish turning-a review. J. Eng. Sci. Technol. Rev. **4**(1), 10–13 (2011)

115. W. König, R. Komanduri, H.K. Toenshoff, G. Ackershott, Machining of hard materials. CIRP Annals **33**(2), 417–427 (1984)

116. T. Ozel, Y. Karpat, A. Srivastava, Hard turning with variable micro-geometry PcBN tools. CIRP Ann. **57**(1), 73–76 (2008)

Chapter 3
Experimentation, Modelling, and Analysis of Machining of Hard Material

Planning and conducting experiments is the key in effective monitoring of system, which leads to success in manufacturing. The traditional approach of experimental study (i.e. one factor at a time, OFAT) requires more number of experiments and consequently consumes more resources. Moreover, the interpretations and analysis that can be made from the experimental data are also limited. Design of experiments (DOE) is a statistical tool, which uses well-planned set of experiments to collect the input–output data. Further, DOE can be used to analyse the experimental data, establish input–output relations, and optimize the process. Figure 3.1 shows the general steps followed in designing a statistical-based experiment.

It is essential to select the process to apply DOE. General guidelines referred in selecting the process for DOE application are mentioned below:

i. Existing process with low productivity.
ii. Required to upgrade or modify the existing process, where changes are required to improve the process efficiency.
iii. The applications especially reverse modelling and online monitoring of the system through feedback and dynamic adjustments. The input–output relations developed via DOE may be used to generate huge amount of data that can be used in reverse modelling through soft computing tools.

The application of DOE needs well-defined responses (output) that can be measured and expressed quantitatively. Further, all parameters influencing the output (response) must be identified along with their operating range. The information on input parameters (variables) and their operating range may be finalized by consulting literature and industry personnel. However, for new process, brainstorming process can be used. It is required to include all parameters initially and screening may be done by applying Plackett and Burman (PB) design. The PB design is able to provide information on important parameters influencing the process. It is important to note that PB design requires very little number of experiments even with huge number of input parameters. The general framework of developing the process or system modelling is presented in Fig. 3.2.

© The Author(s), under exclusive license to Springer Nature Switzerland AG 2020
M. Patel G. C. et al., *Machining of Hard Materials*,
Manufacturing and Surface Engineering,
https://doi.org/10.1007/978-3-030-40102-3_3

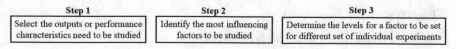

Step 1	Step 2	Step 3
Select the outputs or performance characteristics need to be studied	Identify the most influencing factors to be studied	Determine the levels for a factor to be set for different set of individual experiments

Fig. 3.1 Steps followed while designing a statistical-based experiment

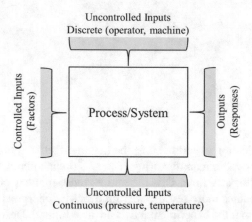

Fig. 3.2 General framework for developing process model

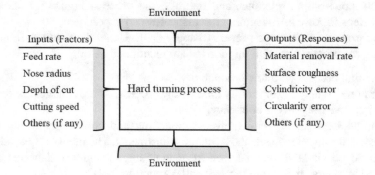

Fig. 3.3 Input–output model for a hard turning process

The choice of important factors influencing the responses is another paramount step to make the process more efficient. In addition, the operating levels chosen for a factor affect the process efficacy. Selecting a wide range operating levels for input parameters (variables) could result in a poor process information, whereas narrow range will result in infeasible solutions for a response [1]. Important to note that the chosen factor must influence directly the output function and should not be treated as a function of other factors or variables. In other words, each input parameter must be independent of other input parameters (Fig. 3.3).

3.1 Selection of Experimental Design

Once the parameters and their operating range are finalized, it is required to select the design matrix and type of design. There are various experimental design matrices available based on full factorial and fractional factorial design (Taguchi, Plackett–Burman, CCD, BBD). Full factorial design will provide detailed insight in terms of accurate estimation of effect of all variables and their interactions. The time and cost incurred in conducting experiments, materials, and labour increase with the increase in number of factors (refer Table 3.1). For example in hard turning process, the parameters such as cutting speed, feed rate, cutting environment (dry, wet, and minimum quantity lubricant), cutting fluid type, chip stability, machine rigidity, tool wear, tool overhang, tool geometry, coating method and coating thickness, work material properties (density, hardness, toughness, strength), material composition, diameter, length and many more will influence the performance characteristics (tool life, form accuracy, material removal rate, surface integrity, productivity and energy consumption). Including all factors in full factorial designs is not practical due to the requirement of large number of experiments. Fractional factorial designs will allow investigators to study the effect of main factors and some interaction factors with less number of experimental trials [2]. Note that although there are many fractional factorial methods available, their required data or sample size, predictive efficiency, assumptions are found approximately similar. The underlying differences are found with reference to the shape of search space or design points (refer Fig. 3.4). This can be clearly viewed with differences in experimental runs for CCD, BBD, and Taguchi method (refer Table 3.1). The final decision on selecting a particular design requires the prerequisite knowledge about factors and their interaction effect on the responses. CCD and BBD are the second-order designs, wherein they possess better distribution of design points (i.e. experiments), minimize error, and estimate precisely the regression function coefficients. The experiments (i.e. design points) in fractional factorial central composite design will include the following [2], (a) set of factorial design, (b) set of centre points (i.e. middle value used in factorial design) and (c) set of axial (α) or star (i.e. points outside the range of factorial design points) points.

Table 3.1 Minimum experiment runs for models operating at fixed three levels of varying factors

Number of factors	Number of experimental runs			
	FFD	CCD	BBD	Taguchi
2	8	9	–	9
3	27	15	13	9 or 27
4	64	25	25	9 or 27
5	125	43	41	27
6	216	77	61	27
7	343	143	85	27
8	512	273	113	27

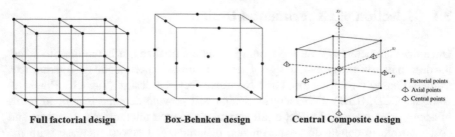

Full factorial design   **Box-Behnken design**   **Central Composite design**

Fig. 3.4 Experimental points of FFD, BBD, and CCD

Distance of axial run, number of experiments to be conducted at centre point, and number of replica are the predefined information required for CCD. Defining the said parameters will decide the rotatability of the composite design. Many researchers use non-rotatable ($\alpha = 1$) central composite design. The rotatable design requires the input variable to be set at five levels, and the geometric nature of practical constraint will limit their usage [1]. Thereby, rotatable design ($\alpha > 1$) should not be chosen until some practical constraint dictates [3]. BBD differs from central composite design with regard to omission of corner points and out-of-boundary points in their experimental design. BBD requires the experiments to be conducted at three levels and few experimental trials as compared to CCD (refer Table 3.1). The experimental runs in BBD will lie at the mid-point of edges of the design space (refer Fig. 3.4). BBD points do not contain all factors at their high or low levels. Thereby, BBD will avoid the corner point experiments which may turn out to yield unsatisfactory results. This design (i.e. BBD) will not provide information regarding the response surface at the extreme corner points (vertices of a cube) [4]. Accurate prediction and optimization might not be possible, when the optimal solution on the response surface will lie at the extreme corner point [1]. Taguchi method posses few drawbacks, such as the quadratic parameters and their significance may not be estimated, although the design allows less experimental trials. Taguchi derived optimal levels for a factor may not be the global solution always; however, it may be used as a starting point for locating the optimal values [5].

In view of the above, the face-centred central composite design has been selected for modelling and optimization of hard turning process. The input–output model of a hard turning process is presented in Fig. 3.3. The inputs and outputs are selected based on expert's opinion from industries, pilot experiments, and literature review. The chemical composition of alloying elements present in the EN 31 steel is shown in Table 3.2. The input parameters and their operating levels selected for modelling of hard turning process are presented in Table 3.3.

Table 3.2 EN 31 chemical composition

Elements	C	Mn	Cr	Co	Si	S	P	Fe
% wt.	0.95–1.2	0.3–0.75	1–1.6	0.025	0.1–0.35	<0.05	<0.04	Balance

Table 3.3 Input factors and levels

Sl. No.	Input factors	Units	Symbol	Levels		
				Low	Medium	High
1	Cutting speed, CS	m/min	A	100	140	180
2	Feed rate, FR	mm/rev	B	0.1	0.2	0.3
3	Depth of cut, DOC	mm	C	0.2	0.3	0.4
4	Nose radius, NR	mm	D	0.4	0.8	1.2

3.2 Workpiece and Tool Material

EN31 (SAE 51100) is a well-known heat-treated bearing steel, popularly used in automotive applications such as bearings, axles, spindle, and moulding dies [6]. EN31 steel possesses excellent hardness, compressive strength, and abrasion resistance. Hence, these materials are found to be difficult to cut with conventional machining processes. Research attempts are made to machine these materials with unconventional machining processes such as electric discharge machining [7, 8], electrochemical machining [9], and plasma arc machining [10]. Although they produce better machining quality, and accuracy, slow material removal rate and high-cost machining restrict their use. Therefore, attempts are made to study the performance of machining quality characteristics during conventional machining of hard turning of EN 31 steel. The chemical composition of alloying elements present in the EN 31 steel is shown in Table 3.2.

Ceramic inserts with three different nose radii (TNGA160404S01525 6050, TNGA160408S01525 6050, and TNGA160412S01525 6050) are used as cutting tool. Machining performance characteristics (material removal rate, circularity error, surface roughness, and cylindricity error) measured while turning the EN 31 steel are presented in Table 3.4.

3.3 Experiment Details

The experiments are planned as per the design matrix of face-centred central composite design. CNC lathe machine (refer Fig. 3.5) is used in machining EN31 steel material with φ 30 mm × 50 mm dimensions. The responses are measured for each experimental condition and the recorded average values (refer Table 3.4).

Machining quality characteristics, such as material removal rate, surface roughness, cylindricity error, and circularity error, are measured and discussed in the following subsections.

Table 3.4 Input–output data of hard turning process

Sl. No.	Input factors				Machining performance characteristics			
	CS (m/min)	FR (mm/rev)	DOC (mm)	NR (mm)	Material removal rate (m³/min)	Surface roughness (μm)	Circularity error (μm)	Cylindricity error (μm)
1	100	0.3	0.4	1.2	0.0240	0.84	1.156	1.769
2	100	0.1	0.4	0.4	0.0080	1.26	1.052	0.473
3	140	0.2	0.2	0.8	0.0112	0.99	0.736	1.080
4	140	0.2	0.3	0.8	0.0168	0.97	0.777	0.686
5	140	0.2	0.3	0.8	0.0168	1.05	0.458	1.004
6	180	0.1	0.2	1.2	0.0072	0.27	0.114	0.867
7	140	0.1	0.3	0.8	0.0084	0.39	0.325	0.542
8	140	0.2	0.3	0.8	0.0168	0.99	0.487	0.909
9	180	0.1	0.4	0.4	0.0144	1.17	0.715	0.585
10	140	0.2	0.4	0.8	0.0224	0.66	0.535	0.658
11	180	0.3	0.4	0.4	0.0432	4.04	0.554	0.890
12	180	0.3	0.2	1.2	0.0216	0.87	0.790	0.792
13	180	0.1	0.4	1.2	0.0144	0.36	0.100	0.625
14	140	0.2	0.3	0.4	0.0168	1.64	0.874	0.930
15	140	0.2	0.3	1.2	0.0168	0.49	0.449	0.601
16	100	0.1	0.2	1.2	0.0040	0.26	0.580	0.661
17	140	0.3	0.3	0.8	0.0252	1.01	0.589	0.870
18	100	0.3	0.2	1.2	0.0120	0.94	0.965	0.740
19	100	0.2	0.3	0.8	0.0120	0.82	1.038	1.937
20	100	0.3	0.2	0.4	0.0120	2.31	0.768	1.391
21	140	0.2	0.3	0.8	0.0168	0.83	0.756	1.143
22	180	0.2	0.3	0.8	0.0216	0.86	1.101	2.148
23	100	0.1	0.4	1.2	0.0080	0.37	0.810	1.158
24	180	0.1	0.2	0.4	0.0072	0.78	1.329	1.750
25	100	0.3	0.4	0.4	0.0240	4.22	0.795	1.426
26	180	0.3	0.4	1.2	0.0432	0.74	1.213	1.758
27	100	0.1	0.2	0.4	0.0040	1.21	1.564	2.045
28	180	0.3	0.2	0.4	0.0216	1.38	1.139	1.576

3.3.1 Material Removal Rate

The machining cost and productivity are directly related with the important quality characteristics, i.e. material removal rate. MRR is computed by utilizing Eq. 3.1.

$$\text{MRR} = \pi \, D_{\text{avg}} \, df \, N \ \text{mm}^3/\text{min} \tag{3.1}$$

Fig. 3.5 Experimental set-up to perform hard turning

$$d = \frac{\text{Initial workpiece diameter} - \text{final workpiece diameter}}{2} \text{ mm}$$

$$D_{\text{avg}} = \frac{\text{Initial workpiece diameter} + \text{final workpiece diameter}}{2} \text{ mm}$$

Terms, D_{avg} signifies the average work piece diameter in mm, d represents the depth of cut in mm, f represents the feed rate in mm/rev, and N corresponds to the spindle speed in rpm.

3.3.2 Surface Roughness

The average vertical deviations measured from the reference surface for a specified length are referred as surface roughness (Ra). The surface roughness on the machined parts at different experimental conditions is measured. For each experimental condition, nine (3 replicates × 3 distinct locations = 9) surface roughness values are measured. The average of 9 surface roughness values obtained for each experimental condition is used. Mitutoyo Surftest SJ301 is used to record the surface roughness values. The cut-off length, maximum traverse speed, and total evaluation length are maintained as 0.8 mm, 0.5 mm/s, and 4 mm, respectively.

3.3.3 Cylindricity and Circularity Error

Circularity error determines the undercut and overcut which occur on the machined surfaces. Maintaining the perfectly round, i.e. cylindrical and circular shapes, straight with no taper on the machined surface is critical. Cylindricity and circularity errors

Fig. 3.6 Samples for
measurement of cylindricity
and circularity error using
3D scanner

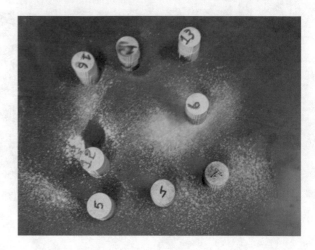

will affect the applications, where close fit in assembly, high level of tolerance, and dimensional accuracy are required. Hence, the applications such as machines, transports, satellite, electric, submarine, aircraft industries need minimum circularity and cylindricity errors. The above-said applications generally use large and heavy cylindrical parts, which are subjected to severe external or internal loads. COMET L3D Tripod, a column-type 3D scanner, is used to measure the form error on the machined samples (refer Fig. 3.6).

3.4   Results and Discussion

The experimental data collected is used to develop nonlinear regression models. This section presents a discussion about developing nonlinear regression models and statistical analysis. The experiments are conducted as per CCD design matrix, and the data is used to develop response-wise regression models.

3.4.1   Response: MRR

The second-order polynomial equation is established for the response MRR, which is expressed as a mathematical function of independent variables (MRR, SR, C_e, and C_E). The regression model for the response, MRR, is presented in Eq. 3.2.

$$\begin{aligned}
MRR = {} & 0.0168 - 0.00012\,A - 0.084\,B - 0.056\,C - 0.000001\,D \\
& - 0.0000001\,A^2 + 0.00001\,B^2 - 0.00001\,C^2 + 0.000001\,D^2 \\
& + 0.0006\,AB + 0.0004\,AC + 0.0000001\,AD + 0.28\,BC \\
& + 0.000001\,BD + 0.000001\,CD
\end{aligned} \tag{3.2}$$

Analysis of variance (ANOVA) is used to examine the statistical adequacy of the developed model (refer Table 3.5). The significance of individual, quadratic, and combined factors is tested for the pre-set 95% confidence level. Model P-value for the response MRR is found to be less than 0.05. Therefore, the model developed for material removal rate is found to be statistically adequate.

The P-values are used to determine the significance of individual, square, and two-term interactions of all variables. The parameters are said to be significant when the corresponding P-values are found to be less than 0.05. The P-values of the variables, namely A (cutting speed), B (feed rate), C (depth of cut), and interactions AB (cutting speed × feed rate), AC (cutting speed × depth of cut), and BC (feed rate × depth of cut), are found to be less than 0.05. Hence, all main factors and two-factor interaction are found to be significant towards the response, MRR. The P-values of square terms of A, B, C, and D are found to be greater than 0.05 (refer Table 3.6), which indicates the existence of strong linear relationship between CS, FR, DOC, and NR with MRR. Feed rate is found to have major impact followed by the depth of cut and cutting speed on MRR. Note that the impact of nose radius is negligible for MRR. The coefficient of multiple determinations is found equal to 0.9959. This signifies that the regression model for the response MRR has good fit, indicating that 99.59% of the variability in the response can be explained by the model.

3.4.2 Response: SR

The second-order regression model is developed for the response, surface roughness and presented in Eq. 3.3.

$$SR = +2.99 - 0.0272\,A + 8.7\,B - 3.2\,C - 1.8\,D + 0.000063\,A^2$$
$$- 3.9\,B^2 + 8.6\,C^2 + 2.04\,D^2 - 0.0119\,AB + 0.0163\,AC$$
$$+ 0.0057\,AD + 23.1\,BC - 8.44\,BD - 7.88\,CD \tag{3.3}$$

Statistical significance of all the terms in Eq. 3.3 is examined for the confidence interval of 95%. Interesting to note that all linear terms (excluding A, cutting speed), and mutual interactions of BC (feed rate × depth of cut), BD (feed rate × nose radius), and CD (depth of cut × nose radius) are statistically significant for the response SR. The P-values of the terms (i.e. A, AB, AC, AD, AA, BB, CC, and DD) are found to be greater than 0.05, indicating that inclusion of these terms in Eq. 3.3 does not contribute much towards SR. All quadratic terms (i.e. square terms) are found insignificant which indicates the existence of strong linear relationship. Nose radius is found to have a major effect followed by feed rate, depth of cut, and cutting speed towards surface roughness. The coefficient of determination value for the regression model is found to be equal to 0.9096, which indicates good fit of polynomial for the response SR. The ANOVA, significance test, and coefficient of determination value

Table 3.5 ANOVA test results for machining quality characteristics

Response		Material removal rate, MRR				Surface roughness, SR			
Source	DF	Adj. SS	Adj. MS	F-value	P-value	Adj. SS	Adj. MS	F-value	P-value
Model	14	0.002508	0.000179	227.41	0.000	22.2019	1.58585	9.35	0.000
Linear	4	0.002249	0.000562	713.88	0.000	16.4464	4.11160	24.23	0.000
Square	4	0.0000001	0.0000001	0.0000	1.000	1.25280	0.31321	1.85	0.180
2-TI	6	0.000259	0.000043	54.71	0.000	4.50270	0.75044	4.42	0.012
Error	13	0.000010	0.000001			2.20610	0.16970		
Lack of fit	10	0.000010	0.000001			2.18010	0.21801	25.15	0.011
Pure error	3	0.0000001	0.0000001			0.02600	0.00867		
Total	27	0.002518				24.1080			

Response		Cylindricity error, C_E				Circularity error, C_e			
Source	DF	Adj. SS	Adj. MS	F-value	P-value	Adj. SS	Adj. MS	F-value	P-value
Model	14	6.37699	0.45550	12.00	0.000	3.02060	0.215757	11.34	0.000
Linear	4	0.74797	0.18699	04.93	0.012	0.70246	0.175616	09.23	0.001
Square	4	2.53459	0.63365	16.70	0.000	0.63071	0.157678	08.29	0.002
2-TI	6	3.09443	0.51574	13.59	0.000	1.68743	0.281238	14.78	0.000
Error	13	0.49336	0.03795			0.24729	0.019022		
Lack of fit	10	0.38266	0.03827	1.04	0.533	0.16021	0.016021	0.55	0.791
Pure error	3	0.11070	0.03690			0.08708	0.087080		
Total	27	6.87036				3.26789			

Table 3.6 Summary of significance test results for machining quality characteristics

Performance characteristics	Correlation coefficient		Process variables	
	All significant and insignificant factor	Removing insignificant factors	Significant factors	Insignificant factors
Material removal rate	0.9959	0.9916	A, B, C, AB, AC, BC	D, AA, BB, CC, DD, AD, BD, CD
Surface roughness	0.9096	0.8123	B, C, D, BC, BD, CD	A, AA, BB, CC, DD, AB, AC, AD
Circularity error	0.9243	0.8428	A, B, D, AA, BB, AB, BD, CD	C, CC, DD, AC, AD, BC
Cylindricity error	0.9282	0.8509	B, D, AA, BB, DD, BC, CD	A, C, CC, AB, AC, AD, BD

have shown that the regression model developed for the response SR is statistically adequate.

The results of the 3D response surface plot of the surface roughness with cutting variables are presented in Fig. 3.7 and are discussed as follows:

1. The value of surface roughness increases, when cutting speed is increased along with FR and DOC. However, surface roughness is observed reducing drastically, when CS is increased with an increase in NR (refer Fig. 3.7a–c). Built-up edge (BUE) formation is negligible beyond the critical levels of the cutting speed. Machining of hardened steel (60 HRC) even at low operating cutting speed of 100 m/min, the formation of built-up edge is not observed during practical experiments. At high cutting speed, plastic behaviour of work material is negligible in the cutting-edge direction, which tends to reduce peak-to-valley height on the surface of the machined part [11]. SR tends to increase with increased values of feed rate and depth of cut. This is expected due to the combined increase in feed rate and depth of cut. The tool–chip interface area increases which induces a large amount of friction and results in increased roughness on the machined surface. Surface finish on the machined part is found to improve with larger nose radii as compared to smaller nose radii. As nose radii are increased, the width of cutting tip edge increases and minimizes the tool profile marks on the machined surface.

2. Combined increase in feed rate and depth of cut resulted in linear increase in the values of surface roughness (refer Fig. 3.7d). This is expected due to the large amount of frictional interface between the work and the tool material at the cutting zone.

3. The surface plots of SR is found to be flat and not much change observed in the response value, when the feed rate is increased along with nose radius (refer Fig. 3.7e).

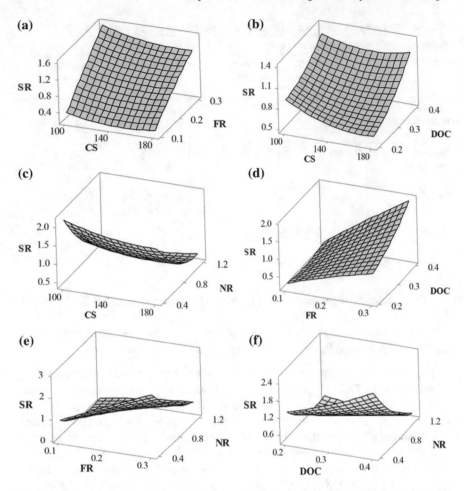

Fig. 3.7 Surface plots of surface roughness versus **a** CS and FR, **b** CS and DOC, **c** CS and NR, **d** FR and DOC, **e** FR and NR, and **f** DOC and NR

4. Figure 3.7f shows the relationship between the nose radius and depth of cut with response surface roughness. As DOC increases, the chip thickness increases as a result of increased friction at the tool–chip interface, which finally leads to rough surface. The available contact area of the tool and work material is small with small nose radii. This could reduce the heat conduction area that favours the rise of temperature near the cutting edge (Fig. 3.8).

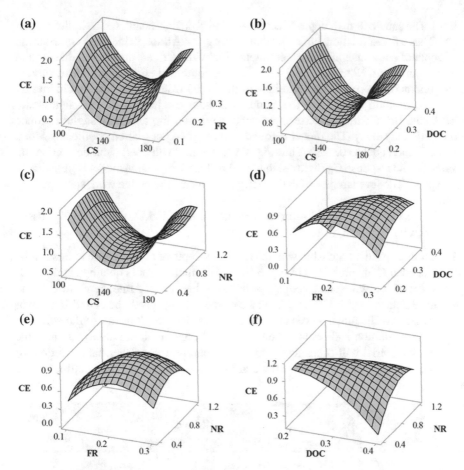

Fig. 3.8 Surface plots of cylindricity error versus **a** CS and FR, **b** CS and DOC, **c** CS and NR, **d** FR and DOC, **e** FR and NR, and **f** DOC and NR

3.4.3   *Response:* C_E

The second-order regression model is developed for the response cylindricity error by utilizing experimental data and presented in Eq. 3.4,

$$
\begin{aligned}
C_E = {}& +12.49 - 0.1694\,A + 6.18\,B + 1.04\,C - 0.56\,D \\
& + 0.000616\,A^2 - 35.1\,B^2 - 18.8\,C^2 - 1.823\,D^2 + 0.0031\,AB \\
& - 0.0175\,AC + 0.00097\,AD + 23.91\,BC \\
& + 2.06\,BD + 8.81\,CD
\end{aligned}
\tag{3.4}
$$

Table 3.6 shows the significant and insignificant terms for the response cylindricity error. The cutting speed and depth of cut are found insignificant at 95% confidence level. The square term of depth of cut is found to be insignificant, which indicates that DOC has a linear relationship with cylindricity error. Although the main contribution of depth of cut is insignificant, their interaction with feed rate and nose radii is found to be significant. Note that the P-value for the combined effect of all linear, square, and interaction terms is found to be less than 0.05 (refer Table 3.5). Excluding all non-contributing terms from the model will make the lack-of-fit term significant. This could result in imprecise input–output relationship and may slightly reduce prediction accuracy. The model is found to have good correlation coefficient with a value found equal to 0.9282. The ANOVA test along with coefficient determination value for the response C_E indicates that the developed regression model is statistically adequate. The few samples of cylindricity error measured on the work specimen are shown in Fig. 3.9.

The results of surface plot for the response and cylindricity error are discussed below (refer Fig. 3.8a–f).

1. Cutting speed relationship with the feed rate, depth of cut, and nose radius for the response C_E is presented in Fig. 3.8a–c. Minimum error is obtained at the mid-values of cutting speed, coupled with low values of feed rate, depth of cut, and nose radius. Beyond the critical value of the cutting speed, the cylindricity error increases. The radial forces in turning process will result in sinusoidal variation in work motion and results in higher C_E. Change in elastic deformation of the work material will occur due to increased cutting forces, thermal distortion, or built-up edge formation with the combination of high feed rate, cutting speed,

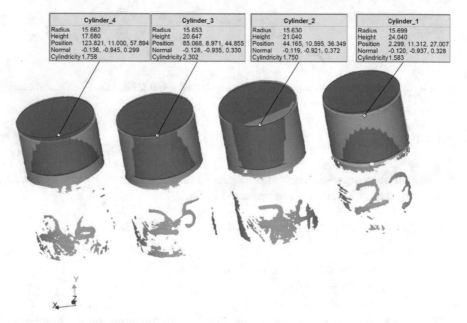

Fig. 3.9 Few test samples of cylindricity error

and depth of cut, and it increases with the radial spindle error. These are the probable reasons for higher cylindricity error in the turned work specimen.

2. Low values of cylindricity error occur at the combination of low values of feed rate, depth of cut, and nose radius (refer Fig. 3.8d–e). It is to be noted that, during part turning, high amount of cutting forces are experienced by the cutting tool at higher feed rate, nose radius, and depth of cut.

3. The contribution of interaction of nose radius with depth of cut (i.e. CD) is more compared with that of other interactions (refer Fig. 3.8f).

3.4.4 Response: C_e

The relationship between cutting variables and circularity error is expressed as mathematical function is presented in Eq. 3.5,

$$C_e = +7.71 - 0.0748\,A - 2.0\,B - 2.36\,C - 2.396\,D + 0.000261\,A^2$$
$$- 19.5\,B^2 - 1.65\,C^2 + 0.059\,D^2 + 0.0275\,AB - 0.01134\,AC$$
$$- 0.00333\,AD + 6.04\,BC + 6.131\,BD + 3.928\,CD \qquad (3.5)$$

The test results of analysis of variance and parameter significance for the response circularity error are presented in Table 3.5. All linear terms (excluding depth of cut) are found insignificant at 95% confidence level. Nose radius is found to have the highest contribution followed by cutting speed and feed rate towards this response. The square terms of depth of cut and nose radius are found insignificant (whose P-values are greater than 0.05), which depicts the relationship of the said terms with the circularity error is linear. Although the term depth of cut is insignificant, their interaction with nose radius (depth of cut × nose radius) is found significant (refer Table 3.6). The feed rate and nose radius (i.e. BD) effects are dominating as compared to the rest of interactions. The value of coefficient of determination for the response circularity error is found equal to 0.9243. Therefore, the model is statistically significant and adequate in making good predictions. Few test samples for measuring circularity error on the machined surface are shown in Fig. 3.10 (Table 3.7).

3.5 Regression Model Validation

The regression models for the responses, namely MRR, SR, C_e, and C_E, are tested for their statistical adequacy through ANOVA, significance test, and coefficient of determination values. The results of statistical analysis are discussed in the previous section. It is also required to validate the regression models developed for their

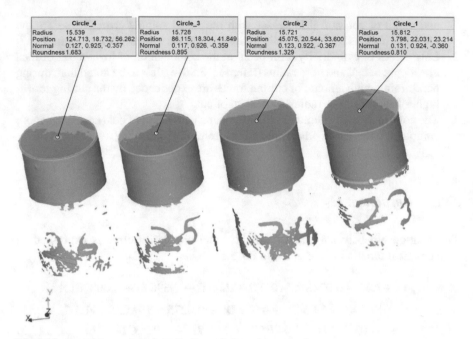

Circle_4		
Radius	15.539	
Position	124.713, 18.732, 56.262	
Normal	0.127, 0.925, -0.357	
Roundness	1.683	

Circle_3		
Radius	15.728	
Position	86.115, 18.304, 41.849	
Normal	0.117, 0.926, -0.359	
Roundness	0.895	

Circle_2		
Radius	15.721	
Position	45.075, 20.544, 33.600	
Normal	0.123, 0.922, -0.367	
Roundness	1.329	

Circle_1		
Radius	15.812	
Position	3.798, 22.031, 23.214	
Normal	0.131, 0.924, -0.360	
Roundness	0.810	

Fig. 3.10 Few test samples of circularity error (roundness error)

prediction accuracy. Prediction accuracy has been tested with the help of test cases, and the results are discussed in this subsection. The experiments are conducted with different combinations of input variables. The input parameter set for each test is generated at random within their operating range (Table 3.8). The prediction accuracy is tested by comparing the response values measured for the test cases with their corresponding values predicted from regression models.

The per cent deviation in predicting response values in ten test cases varies in the range of 2.82–26.94% for MRR, -17.39 to $+17.65\%$ for SR, -13.45 to $+23.33\%$ for C_e, and -13.61 to $+11.11$ for C_E (refer Table 3.8). Important to note that all the responses of the data points predicted on both positive and negative sides of the reference zero line (except MRR). Although MRR model predicted all the data points on the positive sides, those values are still acceptable due to their close to the reference zero line. The average absolute per cent deviation values in prediction of ten test cases are found equal to 12.33% for MRR, 13.17% for SR, 11.05% for C_e, and 8.59% for C_E (refer Table 3.8).

The above discussion shows that the regression models are capable to make good predictions and can be used to determine response values for the known set of input variables without conducting experiments.

Table 3.7 Summary of results of the random test cases

Exp. case	Cutting variables				Experimental values				CCD prediction			
	CS (m/min)	FR (mm/rev)	DOC (mm)	NR (mm)	MRR (m^3/min)	SR (μm)	C_e (μm)	C_E (μm)	MRR (m^3/min)	SR (μm)	C_e (μm)	C_E (μm)
1	123	0.25	0.21	0.4	0.013	1.66	0.92	1.14	0.011	1.885	0.871	1.090
2	114	0.25	0.34	0.4	0.019	2.89	0.82	0.96	0.018	2.752	0.767	1.091
3	146	0.20	0.38	0.8	0.022	1.27	0.54	0.76	0.020	1.069	0.578	0.844
4	104	0.28	0.36	1.2	0.021	0.67	1.25	1.43	0.020	0.777	1.138	1.619
5	103	0.12	0.22	0.8	0.005	0.46	1.10	1.73	0.005	0.527	1.024	1.619
6	174	0.14	0.26	1.2	0.013	0.48	0.41	1.12	0.009	0.435	0.356	0.996
7	153	0.17	0.32	0.8	0.017	0.79	0.54	0.95	0.014	0.669	0.589	1.000
8	126	0.15	0.28	0.4	0.011	1.68	0.85	0.86	0.009	1.384	0.964	0.969
9	138	0.16	0.24	1.2	0.011	0.42	0.37	0.32	0.009	0.451	0.308	0.304
10	174	0.29	0.30	0.4	0.03	2.14	1.10	1.34	0.027	2.512	0.843	1.296

Table 3.8 Summary of results of the random test cases

Exp. case	Per cent deviation in prediction				Absolute per cent deviation in prediction			
	MRR (m³/min)	SR (μm)	C_e (μm)	C_E (μm)	MRR (m³/min)	SR (μm)	C_e (μm)	C_E (μm)
1	13.43	−13.54	5.30	04.34	13.43	13.54	05.30	04.34
2	04.27	04.77	6.47	−13.61	04.27	04.77	06.47	13.61
3	08.77	15.79	−6.96	−11.04	08.77	15.79	06.96	11.04
4	03.60	−15.91	8.99	−13.20	03.60	15.91	08.99	13.20
5	02.82	−14.59	6.95	06.44	02.82	14.59	06.95	06.44
6	26.94	09.34	13.11	11.11	26.94	09.34	13.11	11.11
7	15.69	15.26	−9.08	−5.23	15.69	15.26	09.08	05.23
8	17.92	17.65	−13.45	−12.64	17.92	17.65	13.45	12.64
9	20.72	−7.48	16.83	04.95	20.72	07.48	16.83	04.95
10	09.15	−17.39	23.33	03.30	09.15	17.39	23.33	03.30
Minimum	2.82	−17.39	−13.45	−13.61				
Maximum	26.94	17.65	23.33	11.11				
Average absolute per cent deviation in prediction					12.33	13.17	11.05	08.59

3.6 Concluding Remarks

Statistical modelling and analysis are carried out for the hard turning process. Regression models are developed based on CCD design. Statistical adequacy of all models is tested through ANOVA test. Further, all regression models are tested for their accuracy and practical utility with the help of ten test cases. The following conclusions are drawn from the research work:

1. Feed rate and nose radius (except MRR) are found to have significant contribution towards all the responses. The square terms of depth of cut are found insignificant, as their relationship with all responses is found to be linear. Surface roughness and material removal rate tend to increase linearly with the increase in the values of feed rate and depth of cut. The form errors (circularity and cylindricity error) are found to increase linearly with feed rate and depth of cut. Cutting speed relationship with surface roughness and material removal rate is linear, whereas nonlinear with form errors. It is observed that, as the cutting speed increases, both material removal rate and surface finish of turned part will improve. However, the optimal values of form error are found to be at the middle values of the operating range of cutting speed. Nose radius is the most dominating factor for the responses, namely surface roughness and circularity error.

2. The test results of analysis of variance and parameter significance showed that all responses are statistically adequate. The models are found to have good coefficient of correlation values, indicating good fit of polynomial for the input–output relations.

3. The practical usefulness of the derived regression equations is tested with the help of ten experimental cases. It is to be noted that the experiments are conducted for ten set of input parameters, generated at random, within their respective operating range. The model prediction accuracy is seen to vary both on positive and negative side of reference zero line (except MRR). Moreover, the range of variation in deviation is also found to be within acceptable range. The average absolute per cent deviation in making prediction for each of responses will fall well within 15%. The developed models can be used to predict the test cases for the known set of inputs without conducting experiments.

References

1. G.C.M. Patel, P. Krishna, M.B. Parappagoudar, Squeeze casting process modeling by a conventional statistical regression analysis approach. Appl. Math. Model. 40(15–16), 6869–6888 (2016)
2. D.C. Montgomery, *Design and Analysis of Experiments* (Wiley, New York, 2012)
3. M.B. Parappagoudar, D.K. Pratihar, G.L. Datta, Linear and non-linear statistical modelling of green sand mould system. Int. J. Cast Met. Res. 20(1), 1–13 (2007)
4. S.C. Ferreira, R.E. Bruns, H.S. Ferreira, G.D. Matos, J.M. David, G.C. Brandao, E.P. Da Silva, L.A. Portugal, P.S. Dos Reis, A.S. Souza, W.N.L. Dos Santos, Box-Behnken design: an alternative for the optimization of analytical methods. Anal. Chim. Acta 597(2), 179–186 (2007)
5. I. Mukherjee, P.K. Ray, A review of optimization techniques in metal cutting processes. Comput. Ind. Eng. 50(1–2), 15–34 (2006)
6. L.B. Abhang, M. Hameedullah, Power prediction model for turning EN-31 steel using response surface methodology. J. Eng. Sci. Technol. Rev. 3(1), 116–122 (2010)
7. H.S. Payal, R. Choudhary, S. Singh, Analysis of electro discharge machined surfaces of EN-31 tool steel. J. Sci. Ind. Res. 67, 1072–1077 (2008)
8. N. Faisal, K. Kumar, Optimization of machine process parameters in EDM for EN 31 using evolutionary optimization techniques. Technologies 6(2), 54 (2018)
9. D. Chakradhar, A.V. Gopal, Multi-objective optimization of electrochemical machining of EN31 steel by grey relational analysis. Int. J. Model. Optimization 1(2), 113 (2011)
10. M.K. Das, K. Kumar, T.K. Barman, P. Sahoo, Optimization of process parameters in plasma arc cutting of EN 31 steel based on MRR and multiple roughness characteristics using grey relational analysis. Procedia Mater. Sci. 5, 1550–1559 (2014)
11. K. Bouacha, M.A. Yallese, T. Mabrouki, J.F. Rigal, Statistical analysis of surface roughness and cutting forces using response surface methodology in hard turning of AISI 52100 bearing steel with CBN tool. Int. J. Refract Metal Hard Mater. 28(3), 349–361 (2010)

Chapter 4
Intelligent Modelling of Hard Materials Machining

In the Mid of 1950s, artificial intelligence was emerged to solve practical problems in engineering domain by using tools, developed based on human intelligence. Genetic algorithm 'GA', artificial neural network 'ANN', and fuzzy logic are some AI-based soft computing tools used to predict and assist in control of manufacturing processes. Today, huge money is spent throughout the globe on development of AI technology to assist manufacturing industries. The Artificial Intelligence (AI) has found successful applications in engineering [1] such as selection of materials (i.e. tool and workpiece) [2], tool and vibration monitoring [3, 4], machine and process control [5], fault diagnosis [6], machine design or designing of machine elements [7], production planning and scheduling [8, 9], inspections [10], automatic remeshing using finite element analysis of forging deformation [11], control of welding process [12, 13], motion of aircraft [14, 15], robot [16], laser machining [17], and prediction of properties or responses [18–22].

4.1 Advantages of Artificial Intelligence Over Statistical Methods

The main advantages of AI tools (ANN, fuzzy logic, GAs) over statistical methods are listed below [23, 24]:

1. ANN models do not require prerequisite knowledge about a process.
2. ANN learn with data patterns where the information is incomplete and noisy data.
3. ANN possess good generalization capability.
4. Fuzzy logic models are capable to handle the data with a lot of uncertainty or vagueness to solve the objective functions, which are subjected to practical constraints.

5. GA tools are capable to obtain yield optimal solutions, that enable user to gain cost-effective solution for the industries and customer.
6. GA will conduct heuristic search carried out at many distinct locations in a multidimensional search space to solve the complex optimization task. Optimization of the response functions, even with discrete or complex profile, can be carried out with GA.
7. In many manufacturing systems, there are multiple inputs and outputs affecting the process. Statistical methods (DOE) are capable to model only one response at a time. Statistical DOE may fail to capture the physics or dynamics of process completely, if there is an existence of strong relationship among the various outputs. AI tools can overcome the said problem.
8. Statistical DOE fails to predict simultaneously the multiple outputs, as the data collection, analysis, and models are developed individually for each output. AI tools can overcome the said disadvantages.
9. AI tools predict the multiple responses simultaneously. This will enable the user to adjust the parameters and online monitoring or controlling of the process can be done.
10. Statistical methods cannot be used in predicting the desired set of inputs for the desired output (i.e. reverse mapping). Reverse mapping can be processed or handled effectively with the help of AI tools.
11. AI tools will assist both engineers and managers in decision-making process with the collective information on data analysis, sensitivity, and hypothesis testing.
12. AI tools can incorporate suggestions and recommendation of experts, that could enable to improve the decision quality.
13. AI tools can capture and extract collective information from the data collected through distinguished sources such as experimental design, literature, and surveys. To perform the said task for solving the problem, no complex mathematical formulations or expressions required.

4.2 Neural Networks

Neural networks technique is one of the important AI tools developed to model the real-world problems and are inspired from biology and nature. Learning and adaptation, generalization, parallel computing, simplicity, store and transfer large amount of information are the potential functions of artificial neural networks [25]. The above-mentioned functions will enable the neural networks to use them in wide range of allied disciplines. In 1943, McCulloch and Pitts introduced the first mathematical model with the use of neuron. Later, in 1977, new connection-based models were introduced, namely multi-layer perceptron, trained with different algorithms, adaptive resonance theory and self-organized mapping, and so on. Rapid progress in neural computing and its applications drew much attentions of researchers towards development of many networks such as radial basis function network [26], probabilistic neural network [27], fuzzy neurons and neural network [28–30], recurrent neural networks [31, 32], convolution and deep learning networks [33, 34].

Neural networks can handle and monitor metal cutting process having complex input–output relationship. Neural networks are useful tools, which can be employed in fault diagnosis that could help towards online monitoring of tool wear, chatter vibration, chip breakage, etc. [35], determine the effect of workpiece hardness and tool edge geometry [36], tool-state classification [37], to know the effect of surface roughness and flank wear [38], prediction of tool wear, surface roughness [39], and cutting forces [40]. Table 4.1 shows the recent studies reported in the literature, wherein artificial neural network tools have made performed better predictions. The above applications show that a neural network is a preferable choice due to good learning and extrapolation ability, capable to handle complex nonlinear relationship [41, 42].

Table 4.1 Summary of literature where ANN tool was employed for turning process

Ref.	Models	Prediction of outputs	Architecture (input-hidden-output)	Data		Remarks
				Train	Test	
[43]	BPNN	CF, SR	4-20-4	30	4	NN predictions gave overall 76.4% accuracy
[40]	BPNN	CF, SR, TW	4-8-1; 4-4-1; 4-10-1	54	27	ANN captured the process nonlinearity fully
[44]	BR-NN	CT	3-15-1 for dry 3-20-1 for fluid	22	5	ANN produced better prediction than RSM
[45]	RBF-NN	SR	3-?-1	500	100	RBF-NN produced better predictions
[46]	LM, BR, SCG-NN	SR	3-4-2	34	14	BR-NN offered better prediction of SR
[39]	LM, BR-NN	SR, TW	5-15-2	173	36	LM-NN and BR-NN predictions are found comparable

(continued)

Table 4.1 (continued)

Ref.	Models	Prediction of outputs	Architecture (input-hidden-output)	Data		Remarks
				Train	Test	
[47]	BR-NN	SR	3-17-1	18	9	BR-trained NN offers better predictions
[48]	FFCNN	TW	5-7-1	18	7	The developed model showed better tool wear prediction
[49]	BP, LM, SCG-NN	SR	3-4-1; 3-5-1	21	6	SCG-NN outperformed other neural models for better prediction of SR
[50]	BPNN	SR	3-7-7-1	23	4	ANN model produced better prediction

BP Back-propagation; *BR* Bayesian regularization; *LM* Levenberg–marquardt; *FFCNN* Feedforward-connected neural network; *NN* Neural network; *SCG* Scaled conjugate gradient; *SR* Surface roughness, *CF* Cutting force; *CT* Cutting temperature; and *TW* Tool wear

4.3 Modelling of Hard Turning Process

Important goal of process modelling is to develop a response surface, which is represented by input–output relationship. The parameters such as cutting speed, feed rate, depth of cut, and nose radius significantly influence the responses, namely surface roughness, material removal rate, cylindricity error, and circularity error in hard turning process. The operating levels of input variables in hard turning process are presented in the previous chapter (Table 3.3). The prerequisite steps followed in neural network modelling of hard turning process are presented in Fig. 4.1. From literature (refer Table 4.1), the input variables affecting the hard turning process are identified. Experiments are conducted as per central composite design with four input variables, operating at three levels. The performance characteristics (i.e. circularity error, cylindricity error, surface roughness, and material removal rate) are measured for each experimental run with three replicates. The average values of performance characteristics are recorded and used for further analysis. The experimental data is utilized for developing surface plots (to know the response behaviour when the two input variables are varied simultaneously), testing parameter significance (to know

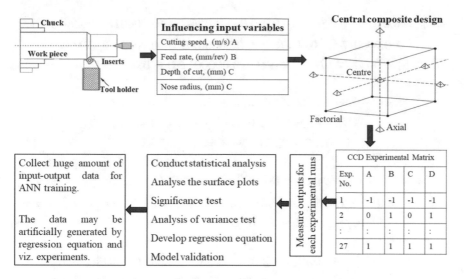

Fig. 4.1 Prerequisite steps in modelling the hard turning process

the contribution level of main and interaction factors), testing model adequacy, and developing regression equations.

4.4   Data Collection for NN Modelling

Neural networks are data-dependent modelling tools, wherein their performances will rely mainly on the data quality and quantity used in model building. The data collection for training and testing of NN-based models, namely BPNN and GA-NN, is discussed below.

4.4.1   Training Data

Huge amount of input–output training data is required to develop NN-based models. NN trained with small data patterns may result in over-fitting, due to the number of connections between the layers with associated weights to be fitted is larger than the data available for training. Therefore, one thousand input–output data is collected, viz. experiments (say, 27) and generated artificially (say, 963) by using regression equations (refer Eqs. 4.1–4.4). This input–output data is supplied to NN in batch mode to optimize the neural network topology (i.e. hidden neurons and layers, synaptic weights, constants of activation function of hidden and output layer, bias, learning rate, and momentum constant).

$$\begin{aligned}
\text{MRR} = \;& 0.0168 - 0.00012A - 0.084B - 0.056C - 0.000001D - 0.0000001A^2 \\
& + 0.00001B^2 - 0.00001C^2 + 0.000001D^2 + 0.0006AB + 0.0004AC \\
& + 0.0000001AD + 0.28BC + 0.000001BD + 0.000001CD \quad\quad (4.1)
\end{aligned}$$

$$\begin{aligned}
\text{SR} = \;& +2.99 - 0.0272A + 8.7B - 3.2C - 1.8D + 0.000063A^2 \\
& - 3.9B^2 + 8.6C^2 + 2.04D^2 - 0.0119AB + 0.0163AC \\
& + 0.0057AD + 23.1BC - 8.44BD - 7.88CD \quad\quad (4.2)
\end{aligned}$$

$$\begin{aligned}
C_E = \;& +12.49 - 0.1694A + 6.18B + 1.04C - 0.56D + 0.000616A^2 - 35.1B^2 \\
& - 18.8C^2 - 1.823D^2 + 0.0031AB - 0.0175AC + 0.00097AD \\
& + 23.91BC + 2.06BD + 8.81CD \quad\quad (4.3)
\end{aligned}$$

$$\begin{aligned}
C_e = \;& +7.71 - 0.0748A - 2.0B - 2.36C - 2.396D + 0.000261A^2 \\
& - 19.5B^2 - 1.65C^2 + 0.059D^2 + 0.0275AB - 0.01134AC \\
& - 0.00333AD + 6.04BC + 6.131BD + 3.928CD \quad\quad (4.4)
\end{aligned}$$

4.4.2 Testing Data

The effectiveness of neural network performance is tested with the help of ten test cases. The test cases include the quality characteristics measured for ten set of input variables. Ten sets of input variables are generated at random within their operating range, and experiments are conducted to measure performance characteristics. The experimental data collected as mentioned above will constitute ten test cases and are not used in training NN (refer Table 4.2). The neural network once trained is allowed to test the prediction accuracy with these ten test examples.

4.5 NN Modelling of Hard Turning Process

NN modelling is carried out based on the collected input–output data derived from both experiments and regression equations. In present work, both forward and reverse modelling is carried out with the acquired input–output data. The input–output model in both forward and reverse direction for hard turning process is presented in Fig. 4.2.

Table 4.2 Summary of results of the random test cases

Exp. Case	Cutting variables				Response values			
	CS (m/min)	FR (mm/rev)	DOC (mm)	NR (mm)	MRR (m³/min)	SR (µm)	C_e (µm)	C_E (µm)
1	123	0.25	0.21	0.4	0.013	1.66	0.92	1.14
2	114	0.25	0.34	0.4	0.019	2.89	0.82	0.96
3	146	0.20	0.38	0.8	0.022	1.27	0.54	0.76
4	104	0.28	0.36	1.2	0.021	0.67	1.25	1.43
5	103	0.12	0.22	0.8	0.005	0.46	1.10	1.73
6	174	0.14	0.26	1.2	0.013	0.48	0.41	1.12
7	153	0.17	0.32	0.8	0.017	0.79	0.54	0.95
8	126	0.15	0.28	0.4	0.011	1.68	0.85	0.86
9	138	0.16	0.24	1.2	0.011	0.42	0.37	0.32
10	174	0.29	0.30	0.4	0.03	2.14	1.10	1.34

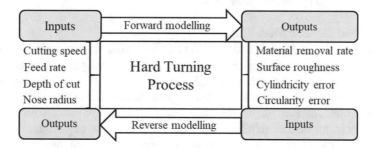

Fig. 4.2 Input–output model for the hard turning process

4.5.1 Forward Modelling

The output of the process is computed in forward modelling. During forward modelling of hard turning process, the machining conditions (i.e. cutting speed, feed rate, depth of cut, and nose radius) are treated as inputs to a network, whereas the performance characteristics (i.e. surface roughness, material removal rate, cylindricity error, and circularity error) are the network outputs.

4.5.2 Reverse Modelling

Reverse modelling aims to predict the inputs for the known set of outputs. In reverse modelling of hard turning process, the performance characteristics (i.e. surface roughness, material removal rate, cylindricity error, and circularity error) are the

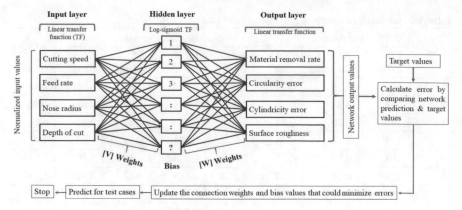

Fig. 4.3 Neural network architecture for hard turning process to predict the responses

inputs, and machining conditions (i.e. cutting speed, feed rate, depth of cut, and nose radius) are treated as outputs of the network.

Forward and Reverse Modelling

Three-layered (input, hidden, and output) neural network structure is used for both forward and reverse modelling. Four neurons in the input layer will represent the machining parameters (i.e. cutting speed, feed rate, depth of cut, and nose radius), and four neurons are assigned to machining performance characteristics (i.e. material removal rate, surface roughness, cylindricity error, and circularity error) in the output layer in forward modelling (refer Fig. 4.3). In reverse modelling, machining performance characteristics will represent the input neurons and machining parameters are represented by neurons in output layer of the network. No universal rule is defined yet to know the choice of hidden layer neurons and is determined after conducting the systematic study. Linear transfer function is used for computing of outputs of the network input layer (refer Eq. 4.5). Logarithmic sigmoid transfer function is employed to determine the network outputs of hidden and output layer (refer Eqs. 4.6 and 4.7).

$$\text{Input layer}: y = mX \tag{4.5}$$

$$\text{Hidden layer}: \ y = 1/(1 + \exp(-aX)) \tag{4.6}$$

$$\text{Output layer}: \ y = 1/(1 + \exp(-bX)) \tag{4.7}$$

Terms, m is estimated after conducting many trials, a and b are the constants of an activation function. X represents the normalized values to input function.

4.6 Back-Propagation Neural Network (BPNN)

Back-propagation algorithm is employed to train the neural network. One thousand (27 experimental and 963 data generated, viz. empirical equation) input–output data sets are used and supplied in batch mode in training NN. The network topology, such as initialized synaptic weights, hidden neurons, learning rate, alpha, constants of activation function, and bias values, is optimized through training.

Supervised learning with one thousand data sets, supplied in batch mode, helps the back-propagation neural network to reduce the error between network predicted and target values. BP algorithm works in two different stages mentioned below

(a) *forward pass computation:* helps in determining the network output
(b) *backward pass computation:* helps to adjust or update the weights that minimize network error.

The mean of squares of all the differences between one thousand network predictions and their corresponding target (i.e. experimental) values is used as an error function in the present work are presented in Eq. 4.8.

$$\text{MSE} = \frac{1}{R \times N} \sum_{i=1}^{R} \sum_{j=1}^{N} \frac{1}{2}\left(T_{ij} - O_{ij}\right)^2 \tag{4.8}$$

where T_{ij} and O_{ij} represent the target and the neural network predicted output, respectively. The weights that act as the connecting strength between the output-hidden layer and input-hidden layer are presented in Eq. 4.9.

$$\Delta w_{jk}(t) = -\eta \frac{\partial E}{\partial w_{jk}}(t) + \alpha \Delta w_{jk}(t-1) \tag{4.9}$$

$$\Delta v_{ij}(t) = -\eta \frac{\partial E}{\partial v_{ij}}(t) + \alpha \Delta v_{ij}(t-1) \tag{4.10}$$

The terms $\partial E / \partial w_{jk}$ and $\partial E / \partial v_{ij}$ are determined by utilizing the chain rule of differentiation. The η and α represent the learning rate and momentum constant.

$$\frac{\partial E}{\partial w_{jk}} = \frac{\partial E}{\partial E_l} \frac{\partial E_l}{\partial E_k} \frac{\partial E_k}{\partial O_{ok}} \frac{\partial O_{ok}}{\partial O_{Ik}} \frac{\partial O_{Ik}}{\partial w_{jk}}$$

$$\frac{\partial E}{\partial v_{ij}} = \frac{\partial E}{\partial E_l} \frac{\partial E_l}{\partial E_k} \frac{\partial E_k}{\partial O_{ok}} \frac{\partial O_{ok}}{\partial O_{Ik}} \frac{\partial O_{Ik}}{\partial H_{oj}} \frac{\partial H_{oj}}{\partial H_{ij}} \frac{\partial H_{ij}}{\partial v_{ij}}$$

The details of parameters of back-propagation neural network are discussed below.

Table 4.3 Selection of
hidden neurons

Reference	Formulae
[56]	$H = 2P + 1$
[57]	$H = \sqrt{(P + M)} + a$
[58]	$H = (P + M)/2$

H Hidden neurons; P Number of input neurons; M Number of
output neurons; and a Constant whose values vary between 1 and
10

4.6.1 Weights

In neural network, the weights contain information with reference to the input signal
of the problem to be solved [51]. Further, the neural network prediction performances
are also relying on initialized weights. The weights initialized with small values may
result in more computation time due to slow convergence, whereas larger weights
may locate local solutions due to network instability. Moreover, equal weights must
not be initialized, since the network may fail to converge and may result in either
stabilizing the error or continuously increasing the error during network training [52,
53]. The present work uses the weight matrices which are generated artificially at
random between the ranges of -1 to $+1$.

4.6.2 Hidden Layers and Neurons

No exact mathematical formulation or rule is derived for the selection of hidden layers
and number of neurons in hidden layer. In general, single hidden layer with enough
training data is sufficient for good generalization and prediction [54]. Although there
are few formulae reported in literature to know the optimal number of neurons in
hidden layer (Table 4.3), those are not globally accepted. Too many neurons in
hidden layer may result in over-fitting problems as a source of poor generalization
and increases computation time [53]. It is difficult to capture the nonlinearity from the
data base collected with the few neurons in the hidden layer. If the number of neurons
in the hidden layer is equal to the training data used, the learning of neural network
will be with poor generalization ability [55]. Therefore, to attain better results, the
ratio of hidden neurons to training data must be maintained equal to 1:10.

4.6.3 Learning Rate and Momentum Constant

To avoid over learning, error vibration, and local optima, the learning rate (η) and
momentum constant (α) need to be chosen carefully. Low values of learning rate
will result in true gradient descent but requires more learning steps to yield better

results. Although the training rates are faster, the solutions getting trapped at local minima are more with high values of learning rate [53]. The momentum constant parameter is used to speed up the search and avoid premature convergence at local minima region. Therefore, η and α values are to be chosen carefully between 0 and 1, after conducting network parameter study [19, 31, 59].

4.6.4 Constants of Activation Function

The main aim of a transfer or activation function is to transform the summation of all weighted values of input neuron to output. Linear transfer function is employed for input layer neuron, which generates the value between 0 and 1. Logarithmic sigmoid activation function is employed for both the hidden and output layer. Few authors had employed constants in activation functions which help in reducing the mean squared error to a minimum value [60, 61]. The systemic study needs to be conducted to determine the optimal value for the constant of transfer function, and it is usually varied between 1 and 10.

4.6.5 Bias

The bias parameter produces the constant output, although it does not receive any input. The purpose of the bias parameter in neural network is to avoid stagnation and control the transfer function during the learning process, which will help the neuron to function more flexible [62]. Noteworthy that the bias value operates with the small value which is varied in the ranges of 0–0.15 [63].

4.7 Genetic Algorithm Neural Network (GA-NN)

Evolutionary GA is one among the popular meta-heuristic stochastic search algorithm and capable to locate near-optimal solution without gradient information. Goldberg developed the computation algorithm (i.e. GA) which is inspired by Darwin's theory applied to natural genetics. It is based on the concept of survival of fittest over many potential solutions. Darwin also stated that the randomly generated initial population of solutions are modified by evaluating the genetic operators (i.e. selection, mutation, and crossover) to hit the global optimal solutions [24]. The discussion on the genetic operators is given below.

4.7.1 Selection

The 'selection' operator will choose the chromosomes from the population, which are necessary for reproduction. Tournament selection is considered as the most popular method, compared to roulette wheel or proportionate, and ranking scheme selection. In tournament selection method, user needs to define initially the size of the tournament and later decides the best solution from the mating pool which are generated at random from the population of solutions. This procedure is repeated several times till the mating pool size is equal to the initialized population size.

4.7.2 Crossover

Crossover operator in GA will share or exchange the properties between the two parents and will result in production of two new offsprings or child solutions. Uniform crossover scheme is paid much attention in recent years as compared to single and multi-point crossover schemes. This could be due to their exhaustive search carried out in the solution space.

4.7.3 Mutation

Reproduction and crossover will generate a wide variety of strings while conducting search in problem or solution space. In GA, the mutation function widens the search space and randomly flips the bits from 1 to 0 and 0 to 1. The mutation rate must be maintained at very low values to avoid stagnation and premature convergence. Note that one complete cycle (iteration) will end after reproduction, crossover, and mutation. It should be noted that the bad solutions from the population are destroyed and only the best solutions are carried from generations to generation.

GA is one among the population-based search methods and used to optimize the parameters of any complex manufacturing process. BP algorithm uses gradient descent approach which always tends to trap at local solution. Thereby, GA replaces the BP algorithm in GA-NN to determine the best parameters (weights, bias, and constants of transfer function) while conducting heuristic search at many spatial locations in a multidimensional space. The parameters are supplied through GA-string, and thus will enable the network to calculate the output. Mean squared error (MSE) is treated as fitness function for the GA-string (refer Eq. 4.11).

$$\text{fitness} = \frac{1}{R \times N} \sum_{i=1}^{R} \sum_{j=1}^{N} \frac{1}{2}\left(T_{ij} - O_{ij}\right)^2 \qquad (4.11)$$

4.8 Results of Forward Mapping

Forward mapping aims to develop the relationship between the cutting variables (i.e. inputs) and machining performance characteristics (i.e. outputs). Three models (CCD, BPNN, and GA-NN) are developed to carry out the forward mapping task (that is predicting the outputs for the known set of inputs).

4.8.1 BPNN

Training of the neural network is carried out using one thousand sets of input–output data, and batch mode of training is adopted. During training, the neural network parameters (hidden neurons, learning rate, momentum constant, activation constant—hidden layer, activation constant—output layer, bias value) are optimized. It is important to note that there are no acceptable universal standards to determine the optimal choice of network parameters. The neural network parameters are decided based on the parameter study (i.e. varying one parameter at a time and keeping the rest of parameters at the fixed level). Note that the optimal choice of values of the network parameters are based on the minimum mean squared error (refer Fig. 4.4a–f). The values of neural network parameters, with a minimum mean squared error values, are presented in Table 4.4.

Ten experimental cases are utilized to check the prediction accuracy of the trained neural network. The mean absolute per cent deviation in prediction of four outputs for the test cases is found equal to 7.13%.

4.8.2 GA-NN

In GA-NN, GA replaces the back-propagation algorithm to train the neural network. The GA-NN prediction performances are related to the determination of optimal network parameters. Therefore, GA parameters (i.e. mutation probability, population size, and generations) are optimized by passing one thousand training data set in batch mode. The minimum mean squared error value corresponds to each genetic operator is selected as an optimal choice of GA-NN (refer Fig. 4.5). The optimal parameters of genetic algorithm neural network obtained at the end of training is presented in Table 4.5.

The performance of GA-NN is tested by utilizing 10 experimental test cases, and the grand average absolute per cent deviation in prediction value is found equal to 6.40%.

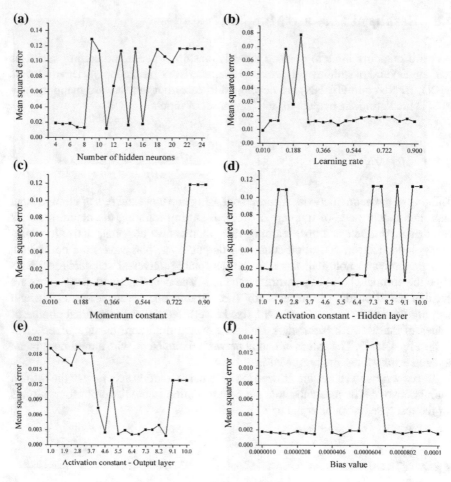

Fig. 4.4 Parameter study of BPNN: **a** MSE versus hidden neurons, **b** MSE versus learning rate, **c** MSE versus momentum constant, **d** MSE versus activation constant-hidden layer, **e** MSE versus activation constant output layer, and **f** MSE versus bias value

Table 4.4 BPNN parameter study

Neural network parameters	Range	Optimal values	MSE
Hidden neurons	4–25	8	0.013220
Learning rate	0.01–0.90	0.01	0.009271
Alpha or momentum constant	0.01–0.90	0.4105	0.003110
Activation constant—hidden layer	1.0–10.0	2.8	0.002225
Activation constant—output layer	1.0–10.0	8.65	0.001837
Bias value	0.000001–0.0001	0.00004555	0.001358

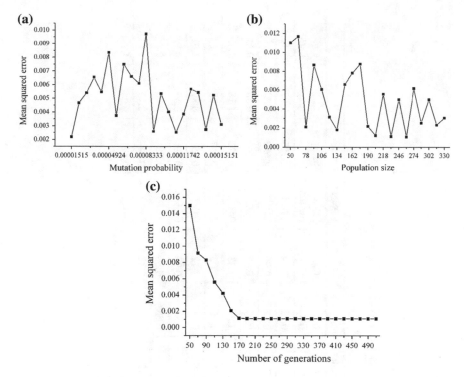

Fig. 4.5 Parameter study of GA-NN: **a** MSE versus mutation probability, **b** MSE versus population size, and **c** MSE versus generations

Table 4.5 GA-NN parameter study

GA parameters	Range	Optimal values	MSE
Mutation probability	0.0001515–0.0001515	0.00001515	0.002188
Population size	50–330	260	0.001092
Generations	50–510	190	0.001092

4.8.3   Summary Results of Forward Mapping

Three models (CCD, BPNN, and GA-NN) are developed and used to predict the responses (such as MRR, C_e, C_E, and SR) of ten experimental cases in hard turning process. The model prediction performances are compared among themselves, and the best model for each response is selected based on average absolute per cent deviation values. Tables 4.6, 4.7, 4.8 and 4.9 presents the summary of results of both BPNN and GA-NN models. The per cent deviation in material removal rate prediction of ten experimental cases is found to vary in the ranges of 0 to 30.77% for CCD, − 20.0 to 15.38% for BPNN, and −15.38 to +13.33% for GA-NN (refer Fig. 4.6a). Interesting to note that, CCD has predicted that data points are found to vary on the

Table 4.6 Summary of results of the random test cases

Exp. Case	Cutting variables					Experimental values				BPNN Prediction			
	CS (m/min)	FR (mm/rev)	DOC (mm)	NR (mm)		MRR (m³/min)	SR (μm)	C_e (μm)	C_E (μm)	MRR (m³/min)	SR (μm)	C_e (μm)	C_E (μm)
1	123	0.25	0.21	0.4		0.013	1.66	0.92	1.14	0.014	1.785	0.971	1.11
2	114	0.25	0.34	0.4		0.019	2.89	0.82	0.96	0.017	2.652	0.78	0.99
3	146	0.20	0.38	0.8		0.022	1.27	0.54	0.76	0.023	1.169	0.58	0.81
4	104	0.28	0.36	1.2		0.021	0.67	1.25	1.43	0.021	0.727	1.18	1.62
5	103	0.12	0.22	0.8		0.005	0.46	1.10	1.73	0.006	0.507	1.08	1.69
6	174	0.14	0.26	1.2		0.013	0.48	0.41	1.12	0.011	0.45	0.39	1.02
7	153	0.17	0.32	0.8		0.017	0.79	0.54	0.95	0.018	0.69	0.59	0.99
8	126	0.15	0.28	0.4		0.011	1.68	0.85	0.86	0.010	1.584	0.88	0.869
9	138	0.16	0.24	1.2		0.011	0.42	0.37	0.32	0.010	0.44	0.33	0.324
10	174	0.29	0.30	0.4		0.030	2.14	1.10	1.34	0.028	2.19	0.943	1.196

Table 4.7 Summary of results of the random test cases

Exp. Case	Per cent deviation in prediction				Absolute per cent deviation in prediction			
	MRR (m³/min)	SR (μm)	C_e (μm)	C_E (μm)	MRR (m³/min)	SR (μm)	C_e (μm)	C_E (μm)
1	−7.69	−7.53	−5.54	2.63	7.69	7.53	5.54	2.63
2	10.53	8.24	4.88	−3.13	10.53	8.24	4.88	3.13
3	−4.55	7.95	−7.41	−6.58	4.55	7.95	7.41	6.58
4	0.00	−8.51	5.60	−13.29	0.00	8.51	5.60	13.29
5	−20.00	−10.22	1.82	2.31	20.00	10.22	1.82	2.31
6	15.38	6.25	4.88	8.93	15.38	6.25	4.88	8.93
7	−5.88	12.66	−9.26	−4.21	5.88	12.66	9.26	4.21
8	9.09	5.71	−3.53	−1.05	9.09	5.71	3.53	1.05
9	9.09	−4.76	10.81	−1.25	9.09	4.76	10.81	1.25
10	6.67	−2.34	14.27	10.75	6.67	2.34	14.27	10.75
Minimum	−20.00	−10.22	−09.26	−13.29				
Maximum	15.38	12.66	14.27	10.75				
Average absolute per cent deviation in prediction					08.89	07.42	06.80	05.41

positive sides of reference zero line for the response MRR. However, neural models (i.e. BPNN and GA-NN) outperformed the CCD model in most of the test cases while predicting the response MRR. Neural model's ability to capture the complete nonlinearity of a process could be the reason of better performance. Figure 4.6b shows that the per cent deviation values in predicting the surface roughness values by the CCD, BPNN, and GA-NN models. The per cent deviation values are seen to vary in the ranges of −17.38 to +17.65% for CCD, −10.22 to +12.66% for BPNN, and −11.45 to +08.86% for GA-NN (refer Tables 3.8, 4.7 and 4.9). It is important to note that, all models have followed a similar pattern of deviation with the data points falling on both side of zero line and also found close to the line (refer Table 4.6). Figure 4.6c compares the performance of three models with reference to per cent deviation in predicting circularity error. The computed deviation from the experimental circularity error is seen to vary in the range of −13.41 to +23.36% for CCD, −9.26 to +14.27% for BPNN, and −5.56 to 13.64% for GA-NN (refer Tables 3.8, 4.7 and 4.9). Noteworthy that, neural network models have outperformed central composite design model in predicting the response circularity error. Figure 4.6d shows that all three models are seen to be comparable in predicting cylindricity error with similar trend. The data points predicted on either side of reference zero line by three models are seen to vary in the range of −13.65 to +11.07% for CCD, −13.29 to +10.75% for BPNN, and −10.49 to +10.71% for GA-NN (refer Tables 3.8, 4.7 and 4.9). GA-NN makes better prediction of cylindricity error as compared to BPNN and CCD. The performance of GA-NN is found to be better than both CCD and BPNN in predicting the response and surface roughness.

Table 4.8 Summary of results of the random test cases

Exp. Case	Cutting variables				Experimental values				GA-NN prediction			
	CS (m/min)	FR (mm/rev)	DOC (mm)	NR (mm)	MRR (m³/min)	SR (μm)	C_e (μm)	C_E (μm)	MRR (m³/min)	SR (μm)	C_e (μm)	C_E (μm)
1	123	0.25	0.21	0.4	0.013	1.66	0.92	1.14	0.015	1.85	0.97	1.13
2	114	0.25	0.34	0.4	0.019	2.89	0.82	0.96	0.018	2.752	0.79	0.98
3	146	0.20	0.38	0.8	0.022	1.27	0.54	0.76	0.023	1.19	0.57	0.83
4	104	0.28	0.36	1.2	0.021	0.67	1.25	1.43	0.02	0.73	1.2	1.58
5	103	0.12	0.22	0.8	0.005	0.46	1.10	1.73	0.005	0.51	1.06	1.71
6	174	0.14	0.26	1.2	0.013	0.48	0.41	1.12	0.012	0.46	0.37	1.00
7	153	0.17	0.32	0.8	0.017	0.79	0.54	0.95	0.019	0.72	0.57	0.98
8	126	0.15	0.28	0.4	0.011	1.68	0.85	0.86	0.012	1.59	0.86	0.88
9	138	0.16	0.24	1.2	0.011	0.42	0.37	0.32	0.011	0.45	0.34	0.33
10	174	0.29	0.30	0.4	0.03	2.14	1.10	1.34	0.026	2.2	0.95	1.21

Table 4.9 Summary of results of the random test cases

Exp. Case	Per cent deviation in prediction				Absolute per cent deviation in prediction			
	MRR (m^3/min)	SR (μm)	C_e (μm)	C_E (μm)	MRR (m^3/min)	SR (μm)	C_e (μm)	C_E (μm)
1	−15.38	−11.45	−5.43	0.88	15.38	11.45	5.43	0.88
2	5.26	4.78	3.66	−2.08	5.26	4.78	3.66	2.08
3	−4.55	6.30	−5.56	−9.21	4.55	6.30	5.56	9.21
4	4.76	−8.96	4.00	−10.49	4.76	8.96	4.00	10.49
5	0.00	−10.87	3.64	1.16	0.00	10.87	3.64	1.16
6	7.69	4.17	9.76	10.71	7.69	4.17	9.76	10.71
7	−11.76	8.86	−5.56	−3.16	11.76	8.86	5.56	3.16
8	−9.09	5.36	−1.18	−2.33	9.09	5.36	1.18	2.33
9	0.00	−7.14	8.11	−3.13	0.00	7.14	8.11	3.13
10	13.33	−2.80	13.64	9.70	13.33	2.80	13.64	9.70
Minimum	−15.38	−11.45	−5.56	−10.49				
Maximum	13.33	08.86	13.64	10.71				
Average absolute per cent deviation in prediction					7.18	7.07	6.05	5.28

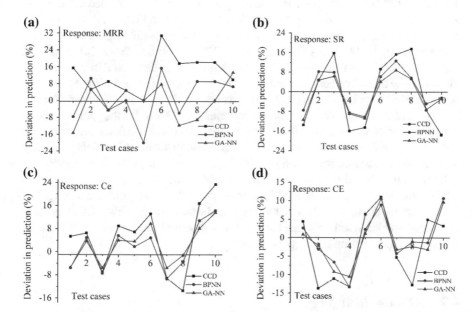

Fig. 4.6 Comparison of the model prediction performances for the responses: **a** MRR, **b** SR, **c** C_e, and **d** C_E

Fig. 4.7 Comparison of
different model prediction
performances considering all
the responses

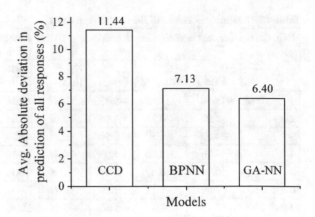

The average absolute per cent deviation values in predicting the different responses
(i.e. MRR, SR, C_e, and C_E) by all the three models (CCD, BPNN, and GA-NN)
are presented in Tables 3.8, 4.7 and 4.9. Neural models (BPNN and GA-NN) have
predicted better response values as compared to statistical central composite design
model. GA-NN has overperformed BPNN in predicting all outputs. This might be
due to the nature of error surface. The aim of training is to reduce the error by
updating weights in each cycle. The mean squared error value obtained at the end of
network training is found equal to 0.001358 for BPNN and 0.001092 for GA-NN.
The grand average absolute per cent deviation value in predicting all responses is
found equal to 11.44% for CCD, 7.13% for BPNN, and 6.40% for GA-NN (refer
Fig. 4.7). Predictions made by GA-NN are found to be better than BPNN and CCD.
This might be due to the complex and multi-modal nature of error surface, where GA-
NN is able to hit global minima. GA conducts heuristic search in a multidimensional
search space at many spatial locations, thus enabling GA to hit the global minima.
Further, GA-NN requires only few tuning parameters (i.e. mutation rate, population
size, and generations) as compared to BPNN. Although the neural models are trained
from the data collected through regression equations, their prediction performance
is found better for all the responses. This might be due to the fact that the neural
models are developed and trained by considering all the responses and might be
able to capture the relative dependency among the outputs. However, CCD model is
developed and analysed only one response at a time and thus unable to capture the
interdependency among the outputs (if any).

4.9 Reverse Mapping

The reverse mapping is conducted to predict the cutting variables (i.e. cutting speed,
feed rate, depth of cut, and nose radius) for the desired set of machining performance
characteristics (i.e. material removal rate, surface roughness, circularity error, and

Table 4.10 BPNN parameter study

Neural network parameters	Range	Optimal values	MSE
Hidden neurons	4–25	12	0.056630
Learning rate	0.01–0.90	0.455	0.008954
Alpha or momentum constant	0.01–0.90	0.495	0.008954
Activation constant–hidden layer	1.0–10.0	3.25	0.004523
Activation constant–output layer	1.0–10.0	6.85	0.001945
Bias value	0.000001–0.0001	0.0000802	0.001454

cylindricity error). Neural network tools (BPNN and GA-NN) are used to develop the reverse process models and their prediction performances are compared among themselves. The transformation matrix might not be invertible always with the derived response equations and hence statistical models (i.e. CCD) may not be able to perform reverse prediction. The data employed in training and testing remains same as that of forward mapping. However, the cutting variables are used as output and machining performance characteristics are treated as inputs of the network.

4.9.1 Back-Propagation NN

The structure of neural network in reverse mapping remains identical to the one employed in forward mapping. The training data consists of cutting variables as network outputs and performance characteristics as inputs to the network. The training data has been supplied in batch mode, and BP algorithm is used to optimize the network parameters. The results of the parameter study of BPNN are presented in Table 4.10. It is observed that the NN parameters are found to be optimal at MSE value equal to 0.001454.

The performance of BPNN in reverse mapping has been tested by using the same ten test cases used in forward mapping. The grand average absolute per cent deviation values in prediction are found equal to 8.66%.

4.9.2 Genetic Algorithm NN

The optimized structure of BPNN is maintained same for GA-NN. Here, GA replaces the back-propagation algorithm to train the genetic algorithm neural network. GA parameters (i.e. mutation probability, population size, and generations) are optimized with the supplied 1000 input–output data sets. The optimized parameters reduce the mean squared error to a minimum value of 0.002165. Table 4.11 shows the results of parameter study of GA-NN. The trained genetic algorithm neural network is tested for

Table 4.11 GA-NN parameter study

GA parameters	Range	Optimal values	MSE
Mutation probability	0.0001515–0.0001515	0.0001106	0.002468
Population size	50–330	260	0.002245
Generations	50–510	510	0.002165

ten random experimental cases and the resulted average absolute per cent deviation in prediction for the cutting variables is found equal to 9.16%.

4.9.3   Summary Results of Reverse Mapping

In reverse mapping, the prediction performances of BPNN and GA-NN models are compared among themselves (refer Tables 4.12, 4.13 and Fig. 4.8a–d). For all responses (i.e. cutting variables), both the models predicted the data points on positive and negative sides with similar trend line. The per cent deviation in prediction by {BPNN and GA-NN} models are seen to vary in the ranges between {−8.77 to +10.57% and −12.28 to +12.20%} for CS, {−16.67 to +13.33% and −17.86 to +15.0%} for FR, {−14.29 to +13.16% and −15.38 to +10.53%} for DOC, and {−15.0 to +10.0% and −20.0 to +8.33%} for NR, respectively (refer Tables 4.14 and 4.15). Interesting to note that, the range of prediction by the BPNN model is found better compared to GA-NN. The average absolute per cent deviation in prediction of cutting variables (CS, FR, NR, and DOC) is found equal to 8.66 for BPNN and 9.16 and GA-NN, respectively (refer Fig. 4.9). Low values of mean squared error of 0.001454 might be reason for better prediction by BPNN, when compared to 0.002165 for GA-NN, respectively. Although GAs conduct heuristic search at many spatial locations in a multidimensional search space, their prediction accuracy is related directly with the nature of error surface and tuning of parameters.

For the industrial relevance in turning process, prediction of set of cutting variables (CS, FR, DOC, and NR) for the desired combination of machining performance characteristics (SR, MRR, C_e, and C_E) is of paramount importance. The results of reverse mapping tools showed that models (i.e. BPNN and GA-NN) can be used for practical utility in industries. Statistical CCD model might fail to conduct reverse prediction due to the transformation matrix to estimate the inverse of combined effect terms in regression equations (refer Eqs. 4.1–4.4) are not invertible always. Further, they fail to predict the multiple outputs simultaneously and capture interdependencies among the variables. Note that, for conducting online monitoring of a process, prediction in both forward and inverse directions are of paramount importance. The reverse mapping predictions are inferior compared to forward mapping. This might be due to the fact that there lies only one output value for the known combination of inputs (i.e. forward mapping), whereas there are multiple sets of inputs for the desired outputs (i.e. reverse mapping) due to interplay among input variables.

Table 4.12 Summary of results of the test cases of BPNN

Exp. Case	Cutting performance characteristics				Experimental cutting values				BPNN prediction			
	MRR (m³/min)	SR (μm)	C_e (μm)	C_E (μm)	CS (m/min)	FR (mm/rev)	DOC (mm)	NR (mm)	CS (m/min)	FR (mm/rev)	DOC (mm)	NR (mm)
1	0.013	1.66	0.92	1.14	123	0.25	0.21	0.4	110	0.28	0.23	0.46
2	0.019	2.89	0.82	0.96	114	0.25	0.34	0.4	124	0.23	0.31	0.43
3	0.022	1.27	0.54	0.76	146	0.2	0.38	0.8	153	0.18	0.33	0.76
4	0.021	0.67	1.25	1.43	104	0.28	0.36	1.2	109	0.32	0.38	1.12
5	0.005	0.46	1.1	1.73	103	0.12	0.22	0.8	100	0.14	0.21	0.84
6	0.013	0.48	0.41	1.12	174	0.14	0.26	1.2	164	0.13	0.29	1.08
7	0.017	0.79	0.54	0.95	153	0.17	0.32	0.8	148	0.19	0.35	0.86
8	0.011	1.68	0.85	0.86	126	0.15	0.28	0.4	135	0.13	0.32	0.43
9	0.011	0.42	0.37	0.32	138	0.16	0.24	1.2	142	0.15	0.27	1.14
10	0.030	2.14	1.1	1.34	174	0.29	0.3	0.4	164	0.31	0.27	0.46

Table 4.13 Summary of results of the test cases of GA-NN

Exp. Case	Cutting performance characteristics				Experimental cutting values				GA-NN Prediction			
	MRR (m³/min)	SR (μm)	C_e (μm)	C_E (μm)	CS (m/min)	FR (mm/rev)	DOC (mm)	NR (mm)	CS (m/min)	FR (mm/rev)	DOC (mm)	NR (mm)
1	0.013	1.66	0.92	1.14	123	0.25	0.21	0.4	108	0.28	0.23	0.48
2	0.019	2.89	0.82	0.96	114	0.25	0.34	0.4	128	0.23	0.31	0.45
3	0.022	1.27	0.54	0.76	146	0.2	0.38	0.8	150	0.17	0.34	0.77
4	0.021	0.67	1.25	1.43	104	0.28	0.36	1.2	112	0.33	0.39	1.16
5	0.005	0.46	1.1	1.73	103	0.12	0.22	0.8	102	0.13	0.2	0.88
6	0.013	0.48	0.41	1.12	174	0.14	0.26	1.2	163	0.15	0.3	1.1
7	0.017	0.79	0.54	0.95	153	0.17	0.32	0.8	146	0.2	0.36	0.88
8	0.011	1.68	0.85	0.86	126	0.15	0.28	0.4	133	0.13	0.3	0.42
9	0.011	0.42	0.37	0.32	138	0.16	0.24	1.2	143	0.17	0.26	1.15
10	0.030	2.14	1.1	1.34	174	0.29	0.3	0.4	165	0.32	0.27	0.45

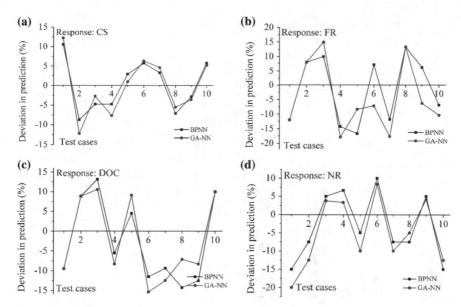

Fig. 4.8 Comparison of the model prediction performances for the responses: **a** CS, **b** FR, **c** DOC, and **d** NR

Table 4.14 Summary of results of the per cent deviation in prediction of test cases—BPNN

Exp. Case	Per cent deviation in prediction				Absolute Per cent deviation in prediction			
	CS (m/min)	FR (mm/rev)	DOC (mm)	NR (mm)	CS (m/min)	FR (mm/rev)	DOC (mm)	NR (mm)
1	10.57	−12.00	−9.52	−15.00	10.57	12.00	9.52	15.00
2	−8.77	8.00	8.82	−7.50	8.77	8.00	8.82	7.50
3	−4.79	10.00	13.16	5.00	4.79	10.00	13.16	5.00
4	−4.81	−14.29	−5.56	6.67	4.81	14.29	5.56	6.67
5	2.91	−16.67	4.55	−5.00	2.91	16.67	4.55	5.00
6	5.75	7.14	−11.54	10.00	5.75	7.14	11.54	10.00
7	3.27	−11.76	−9.37	−7.50	3.27	11.76	9.37	7.50
8	−7.14	13.33	−14.29	−7.50	7.14	13.33	14.29	7.50
9	−2.90	6.25	−12.50	5.00	2.90	6.25	12.50	5.00
10	5.75	−6.90	10.00	−15.00	5.75	6.90	10.00	15.00
Minimum	−8.77	−16.67	−14.29	−15.00				
Maximum	10.57	13.33	13.16	10.00				
Average absolute per cent deviation in prediction					5.67	10.63	9.93	8.42

Table 4.15 Summary of results of the random test cases

Exp. Case	Per cent deviation in prediction				Absolute per cent deviation in prediction			
	CS (m/min)	FR (mm/rev)	DOC (mm)	NR (mm)	CS (m/min)	FR (mm/rev)	DOC (mm)	NR (mm)
1	12.20	−12.00	−9.52	−20.00	12.20	12.00	9.52	20.00
2	−12.28	8.00	8.82	−12.50	12.28	8.00	8.82	12.50
3	−2.74	15.00	10.53	3.75	2.74	15.00	10.53	3.75
4	−7.69	−17.86	−8.33	3.33	7.69	17.86	8.33	3.33
5	0.97	−8.33	9.09	−10.00	0.97	8.33	9.09	10.00
6	6.32	−7.14	−15.38	8.33	6.32	7.14	15.38	8.33
7	4.58	−17.65	−12.50	−10.00	4.58	17.65	12.50	10.00
8	−5.56	13.33	−7.14	−5.00	5.56	13.33	7.14	5.00
9	−3.62	−6.25	−8.33	4.17	3.62	6.25	8.33	4.17
10	5.17	−10.34	10.00	−12.50	5.17	10.34	10.00	12.50
Minimum	−12.28	−17.86	−15.38	−20.00				
Maximum	12.20	15.00	10.53	8.33				
Average absolute per cent deviation in prediction					6.11	11.59	9.97	8.96

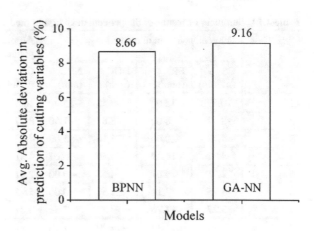

Fig. 4.9 Comparison of different model prediction performances considering all the responses

4.10   Conclusions

Neural network models are developed to establish the input–output relationships of hard turning process in both forward and reverse directions. The conclusions drawn for the present work are discussed below:

1. CCD-based response equations are used to generate huge data for training and optimize the network parameters. The training data has been supplied to the network in batch mode.

2. BPNN and GA-NN models are developed to undergo forward and reverse prediction of hard turning process. Forward modelling aim at prediction of outputs (machining performance characteristics: MRR, SR, C_e, and C_E) for the known set of inputs (cutting variables: CS, FR, DOC, and NR). Reverse modelling is conducted to predict the inputs (cutting variables) for the desired output (machining performance characteristics). Ten test cases generated at random were used to check the prediction performances of both forward and reverse prediction models.

3. In forward mapping, the prediction performances of the developed neural models (BPNN and GA-NN) are compared among themselves and with CCD-derived response equation. Although CCD-derived response equation predicts the data points above the reference zero line (i.e. positive side) for MRR, the neural models predict both on positive and negative sides. Further, the neural models predicted data points found better than CCD model in terms of average absolute per cent deviation in prediction of ten test cases considering all the four responses. In addition, GA-NN outperformed BPNN predictions might be due to the low mean squared error at the termination of training. Moreover, GA conducts heuristic search in multidimensional space at many spatial locations simultaneously, compared to the steepest descent approach of BPNN. The major problem with steepest descent approach is that the probability of getting trapped at local solutions is more as they conduct search in uni-direction.

4. The prime requirements for industries are to know the values of input parameters to obtain the desired output. Neural models (BPNN and GA-NN) are developed to perform the reverse modelling task. Important to note that, the training data and network architecture are maintained approximately similar to that of forward mapping. The data supplied in batch mode to the network optimizes the parameters to attain low mean squared error. The trained networks are tested for practical utility with the ten test cases. BPNN performed better than GA-NN in terms of average absolute per cent deviation in prediction of ten test cases for all cutting variables. The reverse mapping tools are more useful for online control and automate the process.

5. The present work proposes an effective method for conducting forward and reverses mappings of hard turning process using NNs. Thereby, enable to solve the problems related to online monitoring and process automation.

References

1. D.T. Pham, P.T.N. Pham, Artificial intelligence in engineering. Int. J. Mach. Tools Manuf. **39**(6), 937–949 (1999)
2. R.P. Cherian, L.N. Smith, P.S. Midha, A neural network approach for selection of powder metallurgy materials and process parameters. Artif. Intell. Eng. **14**(1), 39–44 (2000)
3. D.F. Hesser, B. Markert, Tool wear monitoring of a retrofitted CNC milling machine using artificial neural networks. Manuf. lett. **19**, 1–4 (2019)

4. W.Y. Chang, C.C. Chen, S.J. Wu, Chatter analysis and stability prediction of milling tool based on zero-order and envelope methods for real-time monitoring and compensation. Int. J. Precis. Eng. Manuf. **20**, 1–8 (2019)
5. D. Luzeaux, Process control and machine learning: Rule-based incremental control. IEEE Trans. Autom. Control **39**(6), 1166–1171 (1994)
6. R. Liu, B. Yang, E. Zio, X. Chen, Artificial intelligence for fault diagnosis of rotating machinery: A review. Mech. Syst. Signal Process. **108**, 33–47 (2018)
7. W. Kacalak, M. Majewski, New intelligent interactive automated systems for design of machine elements and assemblies, in *International Conference on Neural Information Processing* (Springer Berlin Heidelberg, 2012), pp. 115–122
8. S. Nguyen, Y. Mei, M. Zhang, Genetic programming for production scheduling: A survey with a unified framework. Complex Intell. Syst. **3**(1), 41–66 (2017)
9. B. Çaliş, S. Bulkan, A research survey: Review of AI solution strategies of job shop scheduling problem. J. Intell. Manuf. **26**(5), 961–973 (2015)
10. S. Sambath, P. Nagaraj, N. Selvakumar, Automatic defect classification in ultrasonic NDT using artificial intelligence. J. Non-destr. Eval. **30**(1), 20–28 (2011)
11. H. Yano, T. Akashi, N. Matsuoka, K. Nakanishi, O. Takata, N. Horinouchi, An expert system to assist automatic remeshing in rigid plastic analysis. Toyota Tech. Rev. **46**, 87–92 (1997)
12. V. Dey, D.K. Pratihar, G.L. Datta, M.N. Jha, T.K. Saha, A.V. Bapat, Optimization of bead geometry in electron beam welding using a Genetic algorithm. J. Mater. Process. Technol. **209**(3), 1151–1157 (2009)
13. P. Dutta, D.K. Pratihar, Modeling of TIG welding process using conventional regression analysis and neural network-based approaches. J. Mater. Process. Technol. **184**(1–3), 56–68 (2007)
14. A.V. Goncharenko, Several models of artificial intelligence elements for airctaft control, in *2016 4th International Conference on Methods and Systems of Navigation and Motion Control (MSNMC)* (IEEE), pp. 224–227 (2016)
15. L. Gonzalez, G. Montes, E. Puig, S. Johnson, K. Mengersen, K. Gaston, Unmanned Aerial Vehicles (UAVs) and artificial intelligence revolutionizing wildlife monitoring and conservation. Sensors **16**(1), 97 (2016)
16. D.K. Pratihar, K. Deb, A. Ghosh, A genetic-fuzzy approach for mobile robot navigation among moving obstacles. Int. J. Approximate Reasoning **20**(2), 145–172 (1999)
17. T.V. Sibalija, S.Z. Petronic, V.D. Majstorovic, R. Prokic-Cvetkovic, A. Milosavljevic, Multi-response design of Nd: YAG laser drilling of Ni-based superalloy sheets using Taguchi's quality loss function, multivariate statistical methods and artificial intelligence. Int. J. Adv. Manuf. Technol. **54**(5–8), 537–552 (2011)
18. R. Teti, G. Caprino, Prediction of composite laminate residual strength based on a neural network approach. WIT Trans. Inf. Commun. Technol. **6**, WIT Press. www.witpress.com. ISSN 1743-3517
19. P.G. Manjunath, P. Krishna, Prediction and optimization of dimensional shrinkage variations in injection molded parts using forward and reverse mapping of artificial neural networks. Adv. Mater. Res. **463**, 674–678 (2012)
20. M. Patel, P. Krishna, A review on application of artificial neural networks for injection moulding and casting processes. Int. J. Adv. Eng. Sci. **3**(1), 1–12 (2013)
21. M.G. Patel, P. Krishna, M.B. Parappagoudar, Prediction of squeeze cast density using fuzzy logic based approaches. J. Manuf. Sci. Prod. **14**(2), 125–140 (2014)
22. M.G.C. Patel, P. Krishna, M.B. Parappagoudar, Prediction of secondary dendrite arm spacing in squeeze casting using fuzzy logic based approaches. Arch. Foundry Eng. **15**(1), 51–68 (2015)
23. I. Mukherjee, P.K. Ray, A review of optimization techniques in metal cutting processes. Comput. Ind. Eng. **50**(1–2), 15–34 (2006)
24. D.K. Pratihar, Soft computing. Alpha Science International, Ltd. (2007)
25. S. Shanmuganathan, Artificial neural network modelling: An introduction, in *Artificial neural network modelling* (Springer Cham, 2016), pp. 1–14

26. K. Shanmukhi, P.R. Vundavilli, B. Surekha, Modeling of ECDM micro-drilling process using GA-and PSO-trained radial basis function neural network. Soft. Comput. **19**(8), 2193–2202 (2015)
27. R.H.L. Da Silva, M.B. da Silva, A. Hassui, A probabilistic neural network applied in monitoring tool wear in the end milling operation via acoustic emission and cutting power signals. Machining Sci. Technol. **20**(3), 386–405 (2016)
28. P.R. Vundavilli, M.B. Parappagoudar, S.P. Kodali, S. Benguluri, Fuzzy logic-based expert system for prediction of depth of cut in abrasive water jet machining process. Knowl.-Based Syst. **27**, 456–464 (2012)
29. B. Surekha, P.R. Vundavilli, M.B. Parappagoudar, A. Srinath, Design of genetic fuzzy system for forward and reverse mapping of green sand mould system. Int. J. Cast Met. Res. **24**(1), 53–64 (2011)
30. B. Surekha, P.R. Vundavilli, M.B. Parappagoudar, Forward and reverse mappings of the cement-bonded sand mould system using fuzzy logic. Int. J. Adv. Manuf. Technol. **61**(9–12), 843–854 (2012)
31. G.C.M. Patel, A.K. Shettigar, P. Krishna, M.B. Parappagoudar, Back propagation genetic and recurrent neural network applications in modelling and analysis of squeeze casting process. Appl. Soft Comput. **59**, 418–437 (2017)
32. G.C.M. Patel, A.K. Shettigar, M.B. Parappagoudar, A systematic approach to model and optimize wear behaviour of castings produced by squeeze casting process. J. Manuf. Process. **32**, 199–212 (2018)
33. J. Wang, Y. Ma, L. Zhang, R.X. Gao, D. Wu, Deep learning for smart manufacturing: Methods and applications. J. Manuf. Syst. **48**, 144–156 (2018)
34. D. Weimer, B. Scholz-Reiter, M. Shpitalni, Design of deep convolutional neural network architectures for automated feature extraction in industrial inspection. CIRP Ann. **65**(1), 417–420 (2016)
35. M. Rahman, Q. Zhou, G.S. Hong, On-line cutting state recognition in turning using a neural network. Int. J. Adv. Manuf. Technol. **10**(2), 87–92 (1995)
36. J.D. Thiele, S.N. Melkote, Effect of cutting edge geometry and workpiece hardness on surface generation in the finish hard turning of AISI 52100 steel. J. Mater. Process. Technol. **94**, 216–226 (1999)
37. D.E. Dimla Sr., Application of perceptron neural networks to tool state classification in a metal turning operation. Eng. Appl. Artif. Intell. **12**, 471–477 (1999)
38. Y.K. Chou, C.J. Evans, M.M. Barash, Experimental investigation on CBN turning of hardened AISI 52100 steel. J. Mater. Process. Technol. **124**, 274–283 (2002)
39. T. Özel, Y. Karpat, Predictive modeling of surface roughness and tool wear in hard turning using regression and neural networks. Int. J. Mach. Tools Manuf. **45**(4–5), 467–479 (2005)
40. V.N. Gaitonde, S.R. Karnik, L. Figueira, J.P. Davim, Performance comparison of conventional and wiper ceramic inserts in hard turning through artificial neural network modeling. Int. J. Adv. Manuf. Technol. **52**(1–4), 101–114 (2011)
41. K.L. Petri, R.E. Billo, B. Bidanda, A neural network process model for abrasive flow machining operations. J. Manuf. Syst. **17**(1), 52–64 (1998)
42. H.C. Zhang, S.H. Huang, Applications of neural networks in manufacturing: a state-of-the-art survey. Int. J. Product. Res. **33**(3), 705–728 (1995)
43. V.S. Sharma, S. Dhiman, R. Sehgal, S.K. Sharma, Estimation of cutting forces and surface roughness for hard turning using neural networks. J. Intell. Manuf. **19**(4), 473–483 (2008)
44. M. Mia, N.R. Dhar, Prediction of surface roughness in hard turning under high pressure coolant using Artificial Neural Network. Measurement **92**, 464–474 (2016)
45. F.J. Pontes, A.P. de Paiva, P.P. Balestrassi, J.R. Ferreira, M.B. da Silva, Optimization of Radial Basis Function neural network employed for prediction of surface roughness in hard turning process using Taguchi's orthogonal arrays. Expert Syst. Appl. **39**(9), 7776–7787 (2012)
46. M. Mia, N.R. Dhar, Response surface and neural network based predictive models of cutting temperature in hard turning. J. Adv. Res. **7**(6), 1035–1044 (2016)

47. M. Mia, M.H. Razi, I. Ahmad, R. Mostafa, S.M. Rahman, D.H. Ahmed, P.R. Dey, N.R. Dhar, Effect of time-controlled MQL pulsing on surface roughness in hard turning by statistical analysis and artificial neural network. Int. J. Adv. Manuf. Technol. **91**(9–12), 3211–3223 (2017)
48. X. Wang, W. Wang, Y. Huang, N. Nguyen, K. Krishnakumar, Design of neural network-based estimator for tool wear modeling in hard turning. J. Intell. Manuf. **19**(4), 383–396 (2008)
49. I. Asiltürk, M. Çunkaş, Modeling and prediction of surface roughness in turning operations using artificial neural network and multiple regression method. Expert Syst. Appl. **38**(5), 5826–5832 (2011)
50. B.A. Beatrice, E. Kirubakaran, P.R.J. Thangaiah, K.L.D. Wins, Surface roughness prediction using artificial neural network in hard turning of AISI H13 steel with minimal cutting fluid application. Procedia Eng. **97**, 205–211 (2014)
51. S.N. Sivanandam, S.N. Deepa, *Principles of Soft Computing* (Wiley, 2007)
52. H. Kurtaran, B. Ozcelik, T. Erzurumlu, Warpage optimization of a bus ceiling lamp base using neural network model and genetic algorithm. J. Mater. Process. Technol. **169**(2), 314–319 (2005)
53. P.K. Yarlagadda, E.C.W. Chiang, A neural network system for the prediction of process parameters in pressure die casting. J. Mater. Process. Technol. **89**, 583–590 (1999)
54. S. Haykin, *Neural Networks: A Comprehensive Foundation* (Prentice Hall PTR, 1994)
55. S. Rajasekaran, G.V. Pai, *Neural Networks, Fuzzy Logic and Genetic Algorithm: Synthesis and Applications* (with cd) (PHI Learning Pvt. Ltd., 2003)
56. J.Y. Yu, Q. Li, J. Tang, X.D. Sun, Predicting model on ultimate compressive strength of Al2O3-ZrO2 ceramic foam filter based on BP neural network. China Foundry **8**(3), 286–289 (2011)
57. L.H. Jiang, A.G. Wang, N.Y. Tian, W.C. Zhang, Q.L. Fan, BP neural network of continuous casting technological parameters and secondary dendrite arm spacing of spring steel. J. Iron. Steel Res. Int. **18**(8), 25–29 (2011)
58. M.S. Ozerdem, S. Kolukisa, Artificial neural network approach to predict the mechanical properties of Cu–Sn–Pb–Zn–Ni cast alloys. Mater. Des. **30**(3), 764–769 (2009)
59. M.P.G. Chandrashekarappa, P. Krishna, M.B. Parappagoudar, Forward and reverse process models for the squeeze casting process using neural network based approaches. Appl. Comput. Intel. Soft Comput. **2014**, 12 (2014)
60. J.K. Kittur, G.M. Patel, M.B. Parappagoudar, Modeling of pressure die casting process: an artificial intelligence approach. Int. J. Metalcast. **10**(1), 70–87 (2016)
61. G.C.M. Patel, P. Krishna, M.B. Parappagoudar, An intelligent system for squeeze casting process—soft computing based approach. Int. J. Adv. Manuf. Technol. **86**(9–12), 3051–3065 (2016)
62. E. Abhilash, M.A. Joseph, P. Krishna, Prediction of dendritic parameters and macro hardness variation in permanent mould casting of Al-12% Si alloys using artificial neural networks. Fluid Dyn. Mater. Process. **2**, 211–220 (2006)
63. L. Zhang, L. Li, H. Ju, B. Zhu, Inverse identification of interfacial heat transfer coefficient between the casting and metal mold using neural network. Energy Convers. Manag. **51**(10), 1898–1904 (2010)

Chapter 5
Optimization of Machining of Hard Material

In real-life engineering problems, conducting practical experiments and collecting experimental data for analysis and evaluation in order to attain optimal solutions are difficult as compared to data-driven optimization of mathematical functions. In particular, the numerical modelling and simulation process yield solutions and the duration may vary from few seconds to hours depending on the complexity of problems to be solved. Moreover, the solution obtained may or may not be the global optimal solution. Numerical modelling and simulation task can only predict the outputs for set of inputs and needs many try-error runs, which may not yield optimal solutions. On the other hand, optimization tools are capable to locate the global solutions with very less computational efforts, iterations, and time.

Optimization is the process of identifying or determining the most effective solution for an individual or multi-objective function under particular set of constraints. There are two major types of optimization algorithms, namely stochastic and deterministic algorithms. Deterministic search procedure is employed in conventional optimization tools (statistical methods) to move one solution with reference to another, which results in many local solutions. Stochastic algorithms conduct heuristic search in many distinct locations at multi-dimensional search space to locate the optimal solutions. The heuristic search algorithms are generally more suitable to locate the global solutions, although there may be high computation cost due to many local minima or maxima. The technical advantage of stochastic optimization tools is that it will not require exact mathematical formulation and software to optimize the given task. The system works like a black box and does not require high computation efforts (i.e. calculation of derivatives) and time.

Stochastic optimization algorithms are developed based on the idea inspired by natural or biological phenomenon, namely the evolution of species, teacher–learner phenomenon, and behaviour of animals or features of insect colonies. Further, these algorithms have excellent tools for optimization [1]. The technical benefits associated with the structured stochastic search include generating, modifying, and updating the solutions in the wide search space with the collective information carried from the

© The Author(s), under exclusive license to Springer Nature Switzerland AG 2020   103
M. Patel G. C. et al., *Machining of Hard Materials*,
Manufacturing and Surface Engineering,
https://doi.org/10.1007/978-3-030-40102-3_5

previous search locations [2]. Stochastic search-based optimization methods have shown great potential in solving many engineering problems including machining applications. Few structured stochastic optimization techniques found in machining literatures are as follows, genetic algorithm 'GA' [3, 4], particle swarm optimization 'PSO' [5], simulated annealing 'SA' [6], and teaching–learning-based optimization [7]. Although the above stochastic search methods might result in global solutions, they are dependent on the algorithm-specific parameters (crossover, mutation for GA; inertia weight, cognitive, and social leader for PSO; initial and final temperature, and temperature decay rate for SA) [8]. Specific algorithm parameters are to be chosen carefully, while conducting optimization task. No specific standards are developed till date to define the parameters for global solutions. The literature survey shows that the parameter studies have been carried out to determine the algorithm (GA, PSO, ABC, and TLBO) parameters that could produce optimal solutions to sand moulding [9–11], casting [12, 13], machining [14, 15], welding [16], and so on. In addition to the choice of algorithm-specific parameters, the termination criterion for an algorithm to stop optimal search will decide the solution accuracy and computation cost [9]. In general, criteria employed for termination will be either fixed number of generations or total number of cycles or iterations. To know the computational efficiency of different algorithms, several performance indicators, namely convergence plot, solution accuracy, computation time, and cost are used. Some of the popular stochastic algorithms used in manufacturing process optimization are discussed in the following sections.

5.1 Genetic Algorithm

Charles Darwin proposed the concept of evolution, wherein the biological development of species through mating, selection and survival of fittest theory is employed for solving optimization problems [17]. The biologically inspired algorithm is termed as genetic algorithm (GA). GA is a part of the evolutionary algorithm, wherein a set of possible solutions are generated from the population which is included in the chromosomes. GA uses the following steps (refer Fig. 5.1):

Step 1: Initialize the population (or initial candidate solution), wherein many solutions are generated at random for the given problem.

Step 2: Certain percentage of the initialized population is used to modify or create solutions necessary for the next generation. Genetic operator such as reproduction, crossover, and mutation could help to improve the solution.

Reproduction operator forms the mating pool by selecting the potential solutions from the initial population. This operator permits the individual string to be copied for the possible inclusion in the next generation from the current population. The computed string fitness value is treated as the selection criterion in the mating pool. The mating pool thus formed

Fig. 5.1 GA flowchart

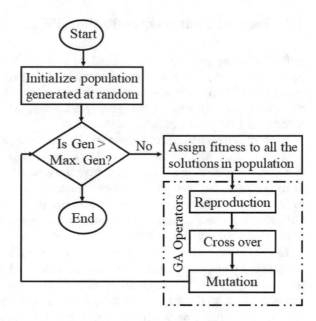

sets the base for the next generation or iteration. GA uses different selection methods, such as proportionate, ranking, steady-state selection, roulette wheel, and tournament [18].

Crossover in biological term refers to the blending of parent chromosomes to generate new offspring chromosome. In GA, crossover aims at generating two new strings (i.e. exchanging the part of parent strings) from the randomly picked two parent strings from the mating pool. There are many crossover schemes (i.e. single point, two point, multi-point, and uniform) are available to generate the offspring strings. Single point crossover does not conduct extensive search but often store the maximum information obtained from the parents and supplied to children solutions. Although uniform crossover is treated as more disruptive among all schemes, they conduct more extensive search in the solution space and storing limited information of the parents in the children solutions.

Although the operators (i.e. reproduction and crossover) generate confounding solutions from different strings, suffer from wide variety of strings to conduct search in the entire search space. Note that GA converges to local solution (not close to global extreme maxima or minima) if the generated initial population in the search space is not good enough. In GA, the mutation operator widens the search space with a simple bit-flip from 1 to 0 and 0 to 1 depending on the small mutation probability. Low mutation probability maintains the already found good solutions, which prevent stagnation and local (i.e. premature) convergence.

GA completes one generation after completing the task of reproduction followed by crossover and mutation. In one complete cycle, the worst solutions from the population (strings) will be destroyed and storing the information of good solutions from generation to generation.

5.2 Particle Swarm Optimization (PSO)

In 1995, Eberhart and Kennedy had introduced the population-based stochastic search optimization tool, which was inspired by social behaviour patterns (i.e. swarming behaviours of bird flocks or schools of fish or swarms of bees) of organisms that live and interact with large number of groups [19]. The idea behind the development and working of PSO algorithm is explained below.

Birds of certain number in a group will be searching randomly the food source in a particular area. Assume that there will be a single piece of food source available in that area. Many birds in a group do not have the information about the exact location of food source, but they pose the approximate distance of the food source during their search. In such a situation, the remaining birds in the group will follow the bird, which is close enough to the food source. Figure 5.2 explains the flowchart illustrating steps involved in PSO.

In PSO algorithm, each bird (i.e. individual or solution) in the search space is referred as a particle. Each particle will move with a definite position and velocity in the search space and the composite of all particles are referred as swarm. Note that all particles are evaluated by the fitness function. Prior to this, initialize the group (i.e. swarm) of particles, which are generated at random and are iteratively updated their positions and velocities over successive generations or iterations. The update of bird position and velocity is the basis of fitness information of flying experience during the food source. In all iteration, each particle is updated based on the fitness information and the cognitive leader (Pbest) and the social leader (Gbest) are selected based on this information. Pbest refers to the personal or particle best position and velocity achieved among all particle solutions evaluated till that time. Gbest is the global best position and velocity solution obtained from the entire population or swarm initialized earlier. The particle velocity is updated with their personal best flying experience and neighbour particle experience in the swarm (refer Eq. 5.1). The particle position alters with change in velocity is presented in Eq. 5.2.

$$\text{New Velocity}: V_i^{K+1} = W \times V_i^k + \text{rand}_1[0, 1]\big[P\text{best}_i^k - P_i^k\big]$$
$$+ \text{rand}_2[0, 1]\big[G\text{best}_i^k - P_i^k\big] \tag{5.1}$$

$$\text{New Position}: P_i^{k+1} = P_i^k + V_i^{k+1} \tag{5.2}$$

The term 'W' refers to the inertia weight and its value has a direct impact on the convergence pattern of PSO. The inertia weight controls the exchange (i.e. trade-off) between the global exploration (wide-range) and local exploitation (near) abilities of the swarm. The inertia weight thus regulates the influence of previous velocities of swarm on the current one. The inertia weight values are varied in the range of zero and one. V_i^k is the current velocity of particle i at iteration k. V_i^{k+1} is the iterative update (new velocity) velocity of particle i, at iteration $k + 1$. $P\text{best}_i^k$ and $G\text{best}_i^k$ represent the best position of particle i, reached until iteration k.

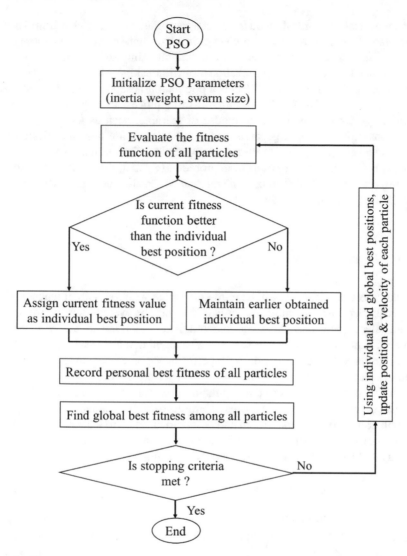

Fig. 5.2 PSO flowchart

5.3 Teaching–Learning-Based Algorithm (TLBO)

Evolutionary (i.e. GA) and swarm intelligence (i.e. PSO) algorithms are the main classification of population-based metaheuristic algorithms. Although the generated solutions by above algorithms are acceptable, however their solution accuracy is dependent on the tuning of algorithm-specific (i.e. mutation rate, crossover, and selection operator for GA; inertia weight, social, and cognitive parameters) and common parameters (i.e. population size, generations, elite size, etc.). Noteworthy that there are no universal standards defined to select appropriate values of algorithm-specific parameters. It is observed that the change in algorithm-specific parameters could result in either increase or decrease of computational time and solution accuracy.

TLBO is introduced as an optimization tool without algorithm-specific parameter in the year 2011 by Rao [20]. TLBO is a population-based optimization algorithm which mimics the behaviour of teaching–learning process in a classroom. Thereby, algorithm works with two general modes of learning, such as teacher phase and learner phase. In TLBO, subjects offered to the learners are treated as design variables, and group of learners is referred as population. The learner results are analogous to fitness value of an optimization problem. In detail, the working principle of teaching–learning-based optimization with teacher and learner phase is discussed in the following section (refer Fig. 5.3).

5.3.1 Teacher Phase

In a classroom, the learners will acquire knowledge with the help of teacher. Here, the teacher is primary attention towards improving the average results of the class in their subject to the best of his/her potential. At iteration i, there are 'n' number of learners (i.e. population size: $k = 1, 2, 3, \ldots n$) and 'm' number of subjects (i.e. design parameters), and $M_{j, i}$ corresponds to the performance of learners in terms of average result in a particular subject 'j'. The overall best performance (i.e. $X_{\text{Total}-k\text{best}, i}$) in all subjects are combined to form the population (i.e. learners), which is analogous to the best learner, $k$best. Note that the teacher refers to highly trained or learned person who will train learners in classroom and best learner refers to teacher in TLBO. The difference between existing average result of each subject and the result associated with each subject of the teacher is computed by using Eq. 5.3.

$$\text{Diff\_Mean}_{j, k, i} = \text{rand}_i[0, \ 1]\left(X_{j, k\text{best}, i} - T_F M_{j, i}\right) \tag{5.3}$$

$X_{j, k\text{best}, i}$ refers to the result of best learner in a subject j. T_F be the teaching factor, whose mean value needs to be updated with either 1 or 2 based on the round-up criteria, presented in Eq. 5.4.

$$\text{Teaching Factor, } T_F = \text{round}[1 + \text{rand}(0, \ 1)\{2 - 1\}] \tag{5.4}$$

In teacher phase, the existing solutions are updated with reference to the solutions obtained from $\text{Diff\_Mean}_{j, k, i}$ according to Eq. 5.5.

$$X_{j, k, i}^* = X_{j, k, i} + \text{Diff\_Mean}_{j, k, i} \tag{5.5}$$

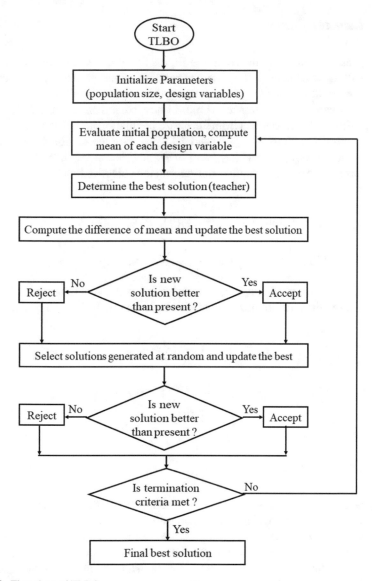

Fig. 5.3 Flowchart of TLBO

$X_{j, k, i}^*$ refers to the updated or modified value of $X_{j, k, i}^*$. The solution of $X_{j, k, i}^*$ is accepted and updated only when they resulted in a better function value. The accepted solutions in teacher phase are used as an input for learner phase.

5.3.2   Learner Phase

Learner phase initiates with the receiving inputs from the teacher phase. Here, the learners will make an attempt to interact at random with other learners in a class to improve their skill or knowledge. Sometimes there is a greater possibility to learn new things, when the other learner has more knowledge than her/him. Let P and Q are the randomly chosen learners from the entire population n, such that $X^*_{\text{Total}-P,i} \neq X^*_{\text{Total}-Q,i}$. Note that terms $X^*_{\text{Total}-P,i}$ and $X^*_{\text{Total}-Q,i}$ are the modified and accepted function values of $X_{\text{Total}-P,i}$ and $X_{\text{Total}-P,i}$ of P and Q learners obtained at the end of teacher phase. If the problem requires optimizing the objective function for minimum value then,

$$X^{**}_{j,P,i} = X^*_{j,P,i} + \text{rand}_i\left(X^*_{j,P,i} - X^*_{j,Q,i}\right), \text{ if } X^*_{\text{Total}-P,i} < X^*_{\text{Total}-Q,i} \qquad (5.6)$$

$$X^{**}_{j,P,i} = X^*_{j,P,i} + \text{rand}_i\left(X^*_{j,Q,i} - X^*_{j,P,i}\right), \text{ if } X^*_{\text{Total}-Q,i} < X^*_{\text{Total}-P,i} \qquad (5.7)$$

Term $X^{**}_{j,P,i}$ is accepted only when they produce better function value. However, the problem requires the optimization of objective functions for high or maximized values then,

$$X^{**}_{j,P,i} = X^*_{j,P,i} + \text{rand}_i\left(X^*_{j,P,i} - X^*_{j,Q,i}\right), \text{ if } X^*_{\text{Total}-Q,i} < X^*_{\text{Total}-P,i} \qquad (5.8)$$

$$X^{**}_{j,P,i} = X^*_{j,P,i} + \text{rand}_i\left(X^*_{j,Q,i} - X^*_{j,P,i}\right), \text{ if } X^*_{\text{Total}-P,i} < X^*_{\text{Total}-Q,i} \qquad (5.9)$$

5.4   JAYA Algorithm

In the year 2016, Rao proposed another metaheuristic algorithm, called JAYA (Sanskrit meaning is victory) which does not require the tuning of algorithm-specific parameters [21, 22]. In general, tuning of algorithm-specific parameters will not only increase the complexity of computation effort and time, but also result in a local optimal solution. Although TLBO is a parameter-free algorithm-specific tool, it needs two function evaluations (i.e. teacher and learner phase) for both teacher and learner phases in any iteration or generation [23, 24]. In view of the above, JAYA algorithm was developed, which will work with a single phase (i.e. only one function evaluation) and result in ease of implementation, less complexity, and less computational time. The unique advantages must balance with the following disadvantages involved in the newly introduced JAYA algorithm [25]. (i) If the population diversity has not been maintained efficiently, then, there is a high convergence rate and may result in local optimal solution, and (ii) there is no standard methodology developed till now to enhance the solution quality in successive generation.

In JAYA, let P be initial number of solutions generated at random subjected to the constrained upper and lower bound of variables. The generated solutions of each variable are updated stochastically by utilizing Eq. 5.10. Let d and F refer to the number of design variables and objective function which may require minimization or maximization. F_best and F_worst tend to be the best and worst solutions corresponding to an objective function. C (i, j, k) will be the value corresponds to jth variable for kth population during ith generation. Therefore, the updated value of C $(i + 1, j, k)$ is given by

$$C(i + 1, j, k) = C(i, j, k) + r(i, j, 1)(C(i, j, b) - |C(i, j, k)|)$$
$$- r(i, j, 2)(C(i, j, w) - |C(i, j, k)|) \qquad (5.10)$$

Terms, b and w will refer to the best and worst solution among the current population. Whereas, i, j, k represent the iteration or generation, design variable, and candidate solution, respectively (i.e. population size, $k = 1, 2, 3, ...n$). $r(i, j, 1)$ and $r(i, j, 2)$ are the random numbers for the jth variable during ith iteration in the range of [0, 1]. Terms $r(i, j, 1)(C(i, j, b) - |C(i, j, k)|)$ will dictate the solution with greater tendency to move close to ideal best solution. However, $- r(i, j, 2)(C(i, j, w) - |C(i, j, k)|)$ term depicts the solution tendency to avoid worst solution. $C(i + 1, j, k)$ is accepted only when it produces better function value. Note that the function values thus accepted after completing the generation are stored and used as input to the next cycle or generation. In short, JAYA algorithm always attempts to reach the best solution (i.e. move close to success) and try to move away from the worst solution (i.e. prevent failure). JAYA algorithm continuously strives to achieve success or victory in gaining the best solution and hence refers its name as JAYA. The working steps of JAYA algorithm are illustrated in Fig. 5.4.

5.5   Modelling and Optimization for Machining Process

Several attempts are made in the past by distinguished researchers to model, analyze, predict, and optimize the parameters (i.e. cutting conditions) for better quality (i.e. surface finish, tool life, and dimensional accuracy) of machining process. DOE (i.e. CCD, FFD, BBD, and Taguchi) method was used to determine the cutting parameters with experimental trials and cost. The following observations are made with reference to different models employed in the literature. Moreover, literature on modelling and optimization of hard turning process is summarized in Table 5.1.

1. Taguchi method: Taguchi method limits the number of practical experiments, which often require controlling the process.
2. Response surface methodology 'RSM' (CCD, BBD, and FFD) is used to establish linear and nonlinear relationship between inputs and outputs.
3. Artificial neural networks (ANN) were used to predict outputs (say, SR, CF, MRR, and so on) for known set of inputs.

Fig. 5.4 JAYA algorithm
flowchart

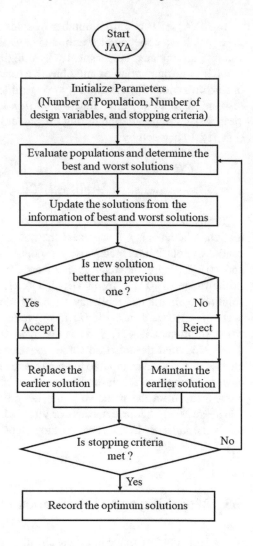

4. PCA: Principal component analysis 'PCA' is employed to transform the multiple objective functions to single objective function (i.e. composite principal component) and this single objective function is optimized. Moreover, the problem of determining the relative importance for individual output is done using PCA.
5. DFA: DFA is generally employed to optimize the multiple objective functions based on criteria such as maximize-the-better, minimize-the-better, and nominal-the-best.
6. GA and PSO: GA and PSO tools are used to optimize the input factors, which will improve the output performances. GA and PSO performances are found better compared to traditional optimization tools. This is due to the heuristic search mechanism carried out at distinct locations in a multi-dimension search space.

Table 5.1 Literature review of modelling and optimization tools applied to hard turning process

Reference	Process outputs	Optimization method	Remarks
[26]	SR	RBF, CCD	RBF neural network designed with CCD found to be effective
[4]	SR	FFD, GA	GA is an effective optimization tool built with the data collected from FFD
[27]	SR, TW	FFD, DFA	DFA produced optimal solutions considering both SR and VB
[28]	SR, CT	Taguchi	Taguchi recommended optimized condition resulted in reduced error
[29]	SR, CF	Taguchi, DFA	The optimized condition resulted with an average of less than 5% variation
[30]	SR	Taguchi	Taguchi recommended optimal levels showed minimum values of surface roughness
[31]	TW, SR, CF, MRR	Taguchi, GRA, DFA, GA	GA produced better optimal solutions compared to other methods
[32]	SR, CF	BBD, DFA	DFA determined optimal conditions are effective based on data collected from BBD
[33]	TW, SR	Taguchi	Good surface finish and minimum tool wear were obtained for Taguchi determined optimized condition
[34]	SR, CF	Taguchi, DFA	Taguchi experimental and DFA optimization tend to improve the performances of hard turning process
[35]	SR, CF	CCD, DFA	DFA determine optimal machining conditions for all output functions
[36]	TW, SR, MRR, MT	ANN, PSO	PSO optimize the conflicting objectives of hard turning process that enable efficient production planning in industries

(continued)

Table 5.1 (continued)

Reference	Process outputs	Optimization method	Remarks
[37]	SR	Taguchi	Taguchi determine the optimal levels for minimum values of surface roughness
[38]	SR, TW	Taguchi	Taguchi recommended optimal levels produced low values of surface roughness
[39]	SR, CF, TW	Taguchi	Taguchi determined optimal conditions resulted in better surface finish with less tool wear
[40]	TW, SR	CCD, SAO	CCD and SAO determined optimal conditions resulted in 9.25% and 8.74% reduction in tool wear and surface roughness
[41]	SR	Taguchi, DFA	DFA determined optimal cutting condition resulted in reduced tool vibrations that could minimize the surface roughness
[42]	SR, MRR	Taguchi, utility concept	Utility approach for simultaneous optimization produces effective results with the relatively ease of implementation
[43]	SR, MH	CCD, DFA	The optimized conditions provide good surface integrity on the machined part
[44]	TL, MT, SR, MRR	CCD, PCA	CCD combined with PCA produced better predictive equation which is found useful for forecasting and optimization
[45]	SR, TL	Taguchi, GA	GA identified optimal conditions resulted in enhanced tool life and surface finish
[46]	CT, SR	Taguchi	Taguchi method determined optimal levels are found useful in optimizing the two responses

(continued)

Table 5.1 (continued)

Reference	Process outputs	Optimization method	Remarks
[47]	CF, SR	CCD, DFA	DFA determined optimized condition resulted in minimum values of cutting forces, surface roughness, and enhanced tool life
[48]	TL, SR, MRR, MT	CCD, PCA	PCA transforms the multiple outputs to a single output function which are essential for optimization
[49]	SR, MRR, DD	Taguchi, GA	GA optimizes the multiple outputs simultaneously which require conflicting requirements
[50]	SR, TW	FFD, DFA	DFA optimized parameters resulted in low machining cost as a result of enhanced tool life and surface finish
[51]	SR, TW, MRR	Taguchi	Taguchi located optimal levels resulted in enhanced surface finish and tool life coupled with better material removal
[52]	SR	FFD, GA, DFA	GA yield higher accuracy to locate the optimal condition compared to DFA

ANN Artificial neural network; *BBD* Box–Behnken design; *CCD* Central composite design; *CF* Cutting forces; *CT* Cutting temperature; *DD* Dimensional deviation; *DFA* Desirability function approach; *FFD* Full factorial design; *GA* Genetic algorithm; *GRA* Grey relational analysis; *MH* Micro-hardness; *MT* Machining time; *MRR* Material removal rate; *PCA* Principal component analysis; *PSO* Particle swarm optimization; *RBF* Radial basis function; *SAO* Sequential approximation optimization; *SR* Surface roughness; *TL* Tool life; *TW* Tool wear

It was observed from the detailed literature review that the soft computing tools (i.e. GA, PSO, and so on) are effective in locating global solutions as compared to statistical or traditional optimization tools. This could be due to the fact that the statistical methods conduct deterministic search with specific set of rules and move one solution with reference to other in a single direction, which may result in local solutions. Conversely, soft computing algorithmic tools will work with certain sets of probabilistic transition rules and conduct heuristic search at many spatial locations in a multi-dimension space could result in greater probability to hit global solutions. Further, not much work is done on the optimization of hard turning process using advanced optimization tools, such as GA, PSO, TLBO, and JAYA.

5.6 Mathematical Formulation for Multi-objective Optimization

In the present work, multiple objective optimization is carried out to optimize the outputs with conflicting nature, which involves maximization of MRR and minimization of SR, C_E, and C_e. Many optimal solutions may be obtained while performing multiple objective optimization and the industry personnel is always keen to know only set of parameter combinations that satisfy all outputs. It is required to obtain one set of input parameter combination, which may be done by converting multi-objective function to a single objective function and optimizing the same. In addition, the optimal solutions vary depending on the relative importance (i.e. weights) assigned to each response. Hence, the (i.e. either maximization or minimization) outputs with conflicting nature are to be modified suitably to a single objective function either to maximize or minimize. The present work is an attempt to optimize the four objective functions simultaneously. Let R_1, (MRR) be the objective function of maximization type, whereas R_2 (SR), R_3 (C_E), and R_4 (C_e) are of minimization type. Five cases with different weights to output are considered. Equal weightage (that is, 25%) is given to all output in first case. Next four cases are considered with highest weightage to one output and equal weightage to all remaining output (that is, 70% to the one with maximum weightage and 10% each for remaining three responses). In each case, the summation of all weight assigned to individual output must be equal to unity. Single objective function is formulated (i.e. global desirability function, D_o) with maximization. The single objective function developed with set of weight fractions is explained below.

$$D_o = \sqrt[4]{y_{MRR}^{w_1} \times y_{SR}^{w_2} \times y_{C_E}^{w_3} \times y_{C_e}^{w_4}} \qquad (5.11)$$

MRR is the output function which requires maximizing the objective function,

$$y_{MRR} = \frac{MRR - MRR_{min}}{MRR_{max} - MRR_{min}}$$

Similarly, SR, C_E, and C_e refer to the objective function of minimization type,

$$y_{SR} = \frac{SR_{max} - SR}{SR_{max} - SR_{min}}, \; y_{C_E} = \frac{C_{Emax} - C_E}{C_{Emax} - C_{Emin}} \; \text{and} \; y_{C_e} = \frac{(C_e)_{max} - C_e}{(C_e)_{max} - (C_e)_{max}}$$

Terms,
Maximum value of surface roughness, SR_{max}
Minimum value of surface roughness, SR_{min}
Maximum value of cylindricity error, C_{Emax}
Minimum value of cylindricity error, C_{Emin}
Maximum value of circularity error, C_{emax}
Minimum value of circularity error, C_{emin}

Maximum value of material removal rate, MRR_{max}
Minimum value of material removal rate, MRR_{min}
The values of objective functions of different responses, such as MRR, SR, C_E, and C_e are determined by using the following equations,

$$
\begin{aligned}
MRR = {}& 0.0168 - 0.00012A - 0.084B - 0.056C - 0.000001D \\
& - 0.0000001A^2 + 0.00001B^2 - 0.00001C^2 + 0.000001D^2 \\
& + 0.0006AB + 0.0004AC + 0.0000001AD \\
& + 0.28BC + 0.000001BD + 0.000001CD
\end{aligned} \tag{5.12}
$$

$$
\begin{aligned}
SR = {}& +2.99 - 0.0272A + 8.7B - 3.2C - 1.8D + 0.000063A^2 \\
& - 3.9B^2 + 8.6C^2 + 2.04D^2 - 0.0119AB + 0.0163AC \\
& + 0.0057AD + 23.1BC - 8.44BD - 7.88CD
\end{aligned} \tag{5.13}
$$

$$
\begin{aligned}
C_e = {}& +7.71 - 0.0748A - 2.0B - 2.36C - 2.396D + 0.000261A^2 \\
& - 19.5B^2 - 1.65C^2 + 0.059D^2 + 0.0275AB - 0.01134AC \\
& - 0.00333AD + 6.04BC + 6.131BD + 3.928CD
\end{aligned} \tag{5.14}
$$

$$
\begin{aligned}
C_E = {}& +12.49 - 0.1694A + 6.18B + 1.04C - 0.56D + 0.000616A^2 \\
& - 35.1B^2 - 18.8C^2 - 1.823D^2 + 0.0031AB - 0.0175AC \\
& + 0.00097AD + 23.91BC + 2.06BD + 8.81CD
\end{aligned} \tag{5.15}
$$

Under control input constraints,
$100 \leq$ Cutting speed, $A \geq 180$
$0.1 \leq$ Feed rate, $B \geq 0.3$
$0.2 \leq$ Depth of cut, $C \geq 0.4$
$0.4 \leq$ Nose radius, $D \geq 1.2$

Terms W_1, W_2, W_3, and W_4 are the weight fractions assigned to output functions of MRR, SR, C_E, and C_e, respectively. Five scenarios are considered and their corresponding fitness value of the function is determined. Note that five scenarios are defined such that the composite values of different fraction of weights in each scenario are equal to unity. Desirability function (DF) approach is used to normalize the fitness function values between zero and one. Zero value of DF refers to completely undesirable response, whereas DF value equal to one corresponds to completely desirable. Note that the normalized highest fitness function value determined from the five scenarios is treated as an optimized condition for multiple conflicting objective functions. Four optimization algorithms (GA, PSO, TLBO, and JAYA) are used to locate the highest fitness function (i.e. global desirability function) values that could improve the machining performance characteristics. Note that the highest value of global desirability function is treated as an optimal condition and the algorithm is treated as the best optimization algorithm for hard turning process.

5.7   Results of Parameter Study of Algorithms (GA, PSO, TLBO, and JAYA)

Parametric study is carried out for all four algorithms, namely GA, PSO, TLBO, and JAYA. The results of parametric study are discussed below.

5.7.1   Genetic Algorithm

The optimum values of GA parameters, namely probability of cross over, probability of mutation, size of population, and generations are determined through parametric study. The results of parameter study for scenario 1 (assigning equal importance to all outputs, i.e. MRR, C_e, SR, and C_E) are shown in Fig. 5.5. The optimal values of genetic algorithm parameters are found equal to 0.85 for P_c, 0.15 for Pm, 100 population size, and 80 number of generations. Five scenarios with different sets of weights are studied to locate the optimal set of cutting variables and machining performance characteristics. The desirability value corresponding to five different case

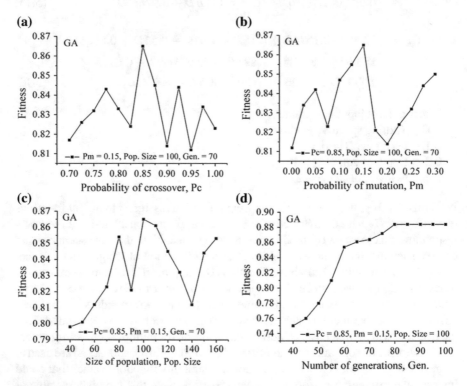

Fig. 5.5 GA parameter study: **a** Pc versus fitness, **b** Pm versus fitness, **c** Pop. size versus fitness, **d** Gen. versus fitness

studies (i.e. scenario 1–5) are found equal to 0.884, 0.890, 0.911, 0.895, and 0.941, respectively (refer Table 5.2). The highest desirability value is found for scenario 5. Hence, scenario 5 could yield the optimal machining performance characteristics subjected to input variable constraints as recommended by GA.

5.7.2 Particle Swarm Optimization

The algorithm-specific parameters (inertia weight, swarm size, and generations) are tuned to attain the highest value of global desirability value. Figure 5.6 shows the parameter study of particle swarm optimization for scenario 1. It has been observed that the maximum desirability value is found when the inertia weight, swarm size, and maximum number of generations are maintained at 0.3, 75, and 70, respectively. The optimal values of parameters which tend to yield the highest desirability value corresponding to five different scenarios are found equal to 0.884, 0.892, 0.939, 0.897, and 0.942, respectively (refer Table 5.2). PSO has shown in scenario 5 (i.e. highest importance assigned to cylindricity error) as the best scenario, yielding better machining performance characteristics.

5.7.3 Teaching–Learning-Based Optimization and JAYA Algorithm

Rao [1, 20] proposed TLBO and JAYA as the specific parameter-less algorithms, to overcome the shortcomings of tuning the parameters and getting stuck at local solutions. Note that the working procedures of both algorithms are different, TLBO uses the two phases (i.e. teacher and learner) and JAYA algorithm locates optimal solutions with only a single phase. Although JAYA and TLBO do not require optimization of algorithm-specific parameters, a proper choice of population size and generations are important. The optimum population size value is found equal to 40 for TLBO and 30 for JAYA. Interesting to note that the maximum number of generations required to yield highest desirability value for scenario 1 was found equal to 40 for JAYA and 60 for TLBO (refer Fig. 5.7). Global desirability values obtained by applying JAYA and TLBO algorithm are found to be equal to {0.893, 0.888}, {0.894, 0.893}, {0.940, 0.930}, {0.903, 0.899}, and {0.943, 0.941}, respectively (refer Table 5.2). The global desirability value for scenario 3 and 5 is comparable. Scenario 5 is recommended by both the algorithms as optimal solution.

Table 5.2 Five scenarios for the computation of optimized conditions of hard turning process

Scenarios	Algorithms	Input variables				Responses				Desirability function value (D_O)
		A	B	C	D	MRR	SR	C_e	C_E	
Scenario 1 (W_1 to W_4 = 0.25)	GA	151.3	0.14	0.37	0.84	0.0137	0.530	0.426	0.648	0.884
	PSO	137.8	0.16	0.35	0.95	0.0133	0.439	0.481	0.742	0.884
	TLBO	165.6	0.12	0.38	0.84	0.0124	0.4063	0.359	0.676	0.888
	JAYA	152.9	0.15	0.35	1.04	0.0136	0.3152	0.371	0.698	0.893
Scenario 2 (W_1 = 0.7, W_2 to W_4 = 0.1)	GA	143.1	0.29	0.35	0.98	0.0266	1.0543	0.625	0.897	0.890
	PSO	134.3	0.29	0.37	0.78	0.0275	1.7854	0.541	0.869	0.892
	TLBO	147.1	0.29	0.35	0.71	0.0277	1.8274	0.532	0.859	0.893
	JAYA	141.1	0.28	0.37	1.03	0.0272	0.9919	0.902	0.894	0.894
Scenario 3 (W_2 = 0.7, W_1, W_3 to W_4 = 0.1)	GA	122.8	0.13	0.27	1.08	0.0069	0.2939	0.415	0.632	0.911
	PSO	146.3	0.15	0.31	1.08	0.0114	0.2851	0.349	0.625	0.939
	TLBO	146.7	0.14	0.28	1.07	0.0091	0.2745	0.295	0.563	0.930
	JAYA	150.4	0.14	0.34	1.08	0.0119	0.2472	0.305	0.589	0.940
Scenario 4 (W_3 = 0.7, W_1, W_2, W_4, = 0.1)	GA	149.9	0.14	0.28	0.93	0.0099	0.3401	0.442	0.821	0.895
	PSO	153.9	0.15	0.29	0.95	0.0107	0.3462	0.447	0.872	0.897
	TLBO	138.9	0.16	0.28	1.05	0.0102	0.3491	0.423	0.675	0.899
	JAYA	145.3	0.18	0.29	1.10	0.0129	0.3995	0.459	0.702	0.903
Scenario 5 (W_4 = 0.7, W_1, W_2, W_3 = 0.1)	GA	132.3	0.13	0.38	0.63	0.0115	0.9301	0.552	0.443	0.941
	PSO	135.1	0.12	0.37	0.76	0.0105	0.5331	0.438	0.426	0.942
	TLBO	142.5	0.11	0.37	0.73	0.0097	0.4711	0.373	0.320	**0.942**
	JAYA	134.7	0.12	0.38	0.82	0.0102	0.4099	0.392	0.397	**0.943**

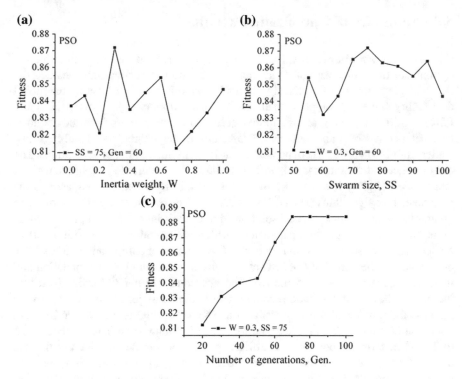

Fig. 5.6 PSO parameter study: **a** W versus fitness, **b** SS versus fitness, and **c** Gen. versus fitness

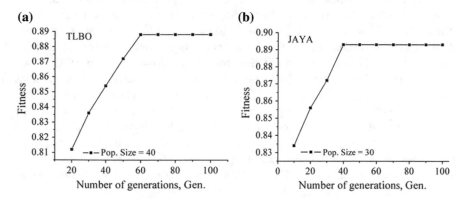

Fig. 5.7 Gen. versus fitness: **a** TLBO and **b** JAYA

5.8   Summary of Optimization Results

Four different algorithms (GA, PSO, TLBO, and JAYA) are used to determine the optimal set of cutting variables yielding best machining performance characteristics. The performance of all the algorithms is tested in terms of locating the global desirability value and computation efforts. The evolutionary algorithms (i.e. GA and PSO) require fine-tuning of algorithm-specific parameters apart from the common parameters (crossover and mutation for GA, and inertia weight, social and cognitive leaders for PSO) such as size of population and maximum number of generations. The performances of evolutionary algorithms are found comparable among themselves and yielded slightly lower values of global desirability value as compared to the parameter-less algorithms (i.e. TLBO and JAYA). The evolutionary algorithms have greater tendency to stuck at local solutions due to their increased complexity, while tuning the specific algorithm parameters. Although evolutionary algorithm performances are comparable, PSO outperformed GA in terms of computation efforts. This occurs due to the simplicity of PSO and requires lesser number of population and number of generations (i.e. 75 and 70) as compared to 100 and 80 for GA. Note that increase in the size of population and maximum number of generations could always increase the number of function evaluations (5250 for PSO and 8000 for GA) and computation time. JAYA algorithm performance is found to be better as compared to TLBO in terms of computation efficiency. This occurs due to JAYA optimized the parameters with minimum number of function evaluations (population size of 30 and generations of $40 = 1200$) as compared to TLBO (i.e. 40 number of population and 60 number of generations $= 2400$). All algorithms have shown that scenario 5 has resulted in highest global desirability value as compared to the rest of the scenarios. JAYA algorithm performance is found marginally better as compared to other algorithms in determining the global desirability value. The global desirability value of JAYA, TLBO, GA, and PSO for scenario 5 is found equal to 0.943, 0.942, 0.941, and 0.942, respectively (refer Table 5.2). Moreover, the number of functions evaluated by JAYA, TLBO, PSO, and GA is found to be equal to 1200, 2400, 5250, and 8000, respectively. JAYA algorithm not only yielded the minimum number of function evaluations but also resulted in highest global desirability value. This is due to simplicity in terms of single phase working of JAYA and no tuning of algorithm-specific parameters. Hence, JAYA algorithm can be used to determine the best set of machining variables (i.e. CS, FR, DOC, and NR) and machining performance characteristics (i.e. MRR, C_e, SR, and C_E) in hard turning process.

5.9   Validation Experiments

Experiments were conducted with the optimized machining conditions as determined by JAYA algorithm. Note that the experiments were conducted in a similar condition as those used in developing models. The results of optimized conditions obtained by applying JAYA algorithm are validated with experimental output values (refer Table 5.3). Note that the average values of three surface roughness measurements, circularity error, and cylindricity error are considered to increase the precision. The JAYA algorithm predicted values are compared with those of experimental output values and the resulted percentage deviation values are found to be equal to 11.2% for surface roughness, 9.88% for circularity error, and 11.2% for cylindricity error. The specimen tested for form error (i.e. circularity and cylindricity error) prepared under optimized condition is presented in Fig. 5.8.

Table 5.3 **Results of** validation experiments for the optimal machining condition

Input variables	JAYA algorithm output values	Experimental output values	Deviation (%)
Cutting speed: 135 m/min	MRR: 0.0102 m^3/min	–	–
Feed rate: 0.12 mm/rev	SR: 0.4099 μm	SR: 0.462 μm	11.2
Depth of cut: 0.38 mm	C_e: 0.392 μm	C_e: 0.435 μm	9.88
Nose radius: 0.8 mm	C_E: 0.397 μm	C_E: 0.447 μm	11.2

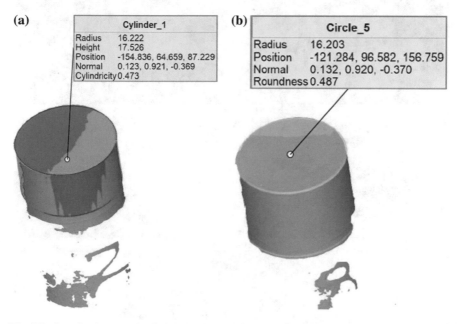

Fig. 5.8 Sample photographs of optimized machining condition: **a** cylindricity error, **b** circularity error

5.10   Tool Wear Studies

The optimal set of cutting parameters obtained by utilizing JAYA algorithm in hard turning process is found to be equal to 0.12 mm/rev, 0.38 mm, and 0.8 mm for FR, DOC, and NR, respectively. The effect of tool life is accessed for the above optimal machining conditions with different tool nose radius (0.4, 0.8, and 1.2 mm). Machining quality characteristics are measured by conducting experiments with input parameters set at optimal condition (refer, Table 5.3). Further, the experiments are conducted by varying only NR to study the effect of NR on tool life. It is to be noted that too low nose radius (that is, NR: 0.4 mm) resulted in chipping of the cutting tool and break (refer Fig. 5.9). High temperature is generated with low nose radius (smaller cutting tip edge) due to narrow tool-chip contact area and increased chip thickness. The cutting tool with small nose radius (NR < 0.4 mm) will not be capable to withstand high cutting temperature and cutting forces. Further, the cutting tool inserts with NR value equal to 0.8 mm and 1.2 mm have not shown much significant

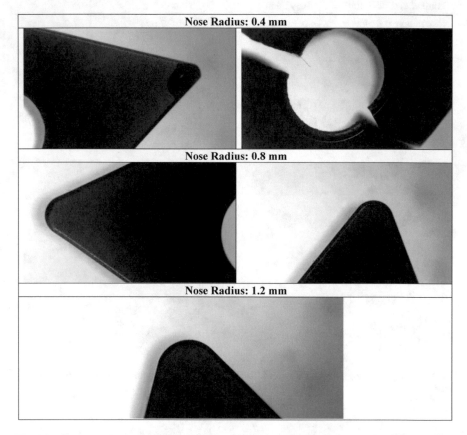

Fig. 5.9   Cutting tool insert performances with different nose radius (0.4 mm, 0.8mm and 1.2 mm)

difference in performance (refer Fig. 5.9). The optimum nose radius is fairly good and in agreement with the experimental studies on hard turning [53, 54]. Cutting tools with a larger nose radius (i.e. large width of cutting tip edge) are recommended to maintain good structural strength of tool (edge geometry) and surface finish on work material [53]. Note that the tool nose radius need not be increased much, as it critically influences the machining surface quality and dimensional accuracy [55].

5.11   Conclusions

Hard materials are extensively used in industries and processed through machining industries. It is required to know the best set of cutting variables that will result in good machining performance along with reduction in cost. Various optimization algorithms are used to optimize hard turning process and the following conclusions are drawn:

1. CCD-based response surface methodology is used to develop empirical input–output relationships in hard turning process. These response equations are used as an objective function for optimizing the process.
2. Four soft computing-based algorithms are (GA, PSO, TLBO, and JAYA) are used as optimization tool to determine the best set of cutting variables (CS, FR, DOC, and NR) that could simultaneously optimize the multiple output variables (minimize: C_E, C_e, and SR; maximize: MRR).
3. Five scenarios (equal importance to all outputs, and maximum importance to one output after keeping the rest at low values) are considered to determine the most suitable optimizing conditions for hard turning process.
4. Scenario 5 (i.e. maximum importance to cylindricity error) is recommended by all the optimization algorithms as their corresponding desirability value is found close to unity.
5. Evolutionary algorithms (GA and PSO) require fine-tuning of algorithm-specific parameters to attain global desirability value. The detailed parameter study is conducted to avoid local minima. The GA and PSO results are found comparable, but GA requires maximum number of function evaluations which consumes more computation time.
6. The JAYA and TLBO are algorithm-specific parameter-less optimization tools and outperformed evolutionary algorithm s, namely GA and PSO in determining global desirability value and computation time. Although JAYA algorithm results are comparable with TLBO in terms of achieving global desirability value, it requires comparatively lesser number of function evaluations.
7. The results obtained through JAYA algorithm are tested by conducting the confirmation experiments. The per cent deviation value is found to be equal to 11.2% for SR, 9.88% for C_e, and 11.2% for cylindricity error.

8. The tailormade results are useful to the machining industries to achieve the best results in machining quality and economics in machining. The methodology and results are useful for the machining industries.

References

1. R.V. Rao, A. Saroj, An elitism-based self-adaptive multi-population Jaya algorithm and its applications. Soft. Comput. **23**(12), 4383–4406 (2019)
2. A.P. Engelbrecht, *Computational Intelligence: An Introduction* (Wiley, 2007)
3. H. Ganesan, G. Mohankumar, Optimization of machining techniques in CNC turning centre using genetic algorithm. Arab. J. Sci. Eng. **38**(6), 1529–1538 (2013)
4. D. Singh, P. Venkateswara Rao, Optimization of tool geometry and cutting parameters for hard turning. Mater. Manuf. Process. **22**(1), 15–21 (2007)
5. A. Panda, S.R. Das, D. Dhupal, Surface roughness analysis for economical feasibility study of coated ceramic tool in hard turning operation. Process Integr. Optim. Sustain. **1**(4), 237–249 (2017)
6. M. Mia, N.R. Dhar, Modeling of surface roughness using RSM, FL and SA in dry hard turning. Arab. J. Sci. Eng. **43**(3), 1125–1136 (2018)
7. M. Mia, G. Królczyk, R. Maruda, S. Wojciechowski, Intelligent optimization of hard-turning parameters using evolutionary algorithms for smart manufacturing. Materials **12**(6), 879 (2019)
8. R.V. Rao, Teaching-learning-based optimization algorithm, in *Teaching learning based optimization algorithm* (Springer, Cham, 2016), pp. 9–39. https://doi.org/10.1007/978-3-319-22732-0_2
9. B. Surekha, L.K. Kaushik, A.K. Panduy, P.R. Vundavilli, M.B. Parappagoudar, Multi-objective optimization of green sand mould system using evolutionary algorithms. Int. J. Adv. Manuf. Technol. **58**(1–4), 9–17 (2012)
10. G.R. Chate, G.C.M. Patel, A.S. Deshpande, M.B. Parappagoudar, Modeling and optimization of furan molding sand system using design of experiments and particle swarm optimization. Proc. Inst. Mech. Eng. Part E: J. Process Mech. Eng. **232**(5), 579–598 (2018)
11. G.R. Chate, G.M. Patel, S.B. Bhushan, M.B. Parappagoudar, A.S. Deshpande, Comprehensive modelling, analysis and optimization of furan resin-based moulding sand system with sawdust as an additive. J. Brazilian Soc. Mech. Sci. Eng. **41**(4), 183 (2019). https://doi.org/10.1007/s40430-019-1684-0
12. G.C.M. Patel, A.K. Shettigar, M.B. Parappagoudar, A systematic approach to model and optimize wear behaviour of castings produced by squeeze casting process. J. Manuf. Process. **32**, 199–212 (2018)
13. G.C.M. Patel, P. Krishna, M.B. Parappagoudar, P.R. Vundavilli, S.B. Bhushan, Squeeze casting parameter optimization using swarm intelligence and evolutionary algorithms, in *Critical developments and applications of swarm intelligence* (IGI Global, 2018), pp. 245–270
14. R.V. Rao, D.P. Rai, J. Balic, A multi-objective algorithm for optimization of modern machining processes. Eng. Appl. Artif. Intell. **61**, 103–125 (2017)
15. D.M. Babu, S.V. Kiran, P.R. Vundavilli, A. Mandal, Experimental investigations and multi-response optimisation of wire electric discharge machining of hypereutectic Al-Si alloys. Int. J. Manuf. Res. **11**(3), 221–237 (2016)
16. K. Srinivas, P.R. Vundavilli, M.M. Hussain, Optimization of weld-bead parameters of plasma arc welding using GA and IWO, in *Advances in manufacturing technology* (Springer Singapore, 2019), pp. 17–24
17. O. Kramer, *Genetic algorithm essentials*, vol. 679 (Springer, Singapore, 2017)
18. N.M. Razali, J. Geraghty, Genetic algorithm performance with different selection strategies in solving TSP, in *Proceedings of the world congress on engineering*, vol. 2(1) (International Association of Engineers, Hong Kong, 2011), pp. 1–6

19. J. Kennedy, Swarm intelligence. in *Handbook of nature-inspired and innovative computing* (Springer, Boston, MA, 2006), pp. 187–219. https://doi.org/10.1007/0-387-27705-6_6

20. R.V. Rao, V.J. Savsani, D.P. Vakharia, Teaching–learning-based optimization: a novel method for constrained mechanical design optimization problems. Comput. Aided Des. **43**(3), 303–315 (2011)

21. R.V. Rao, Teaching-learning-based optimization algorithm, in *Teaching learning-based optimization algorithm* (Springer, Cham, 2016), pp. 9–39

22. R.V. Rao, Jaya: A simple and new optimization algorithm for solving constrained and unconstrained optimization problems. Int. J. Ind. Eng. Comput. **7**, 19–34 (2016)

23. R.V. Rao, D.P. Rai, Optimisation of welding processes using quasi-oppositional-based Jaya algorithm. J. Exp. Theor. Artif. Intell. **29**(5), 1099–1117 (2017)

24. R.V. Rao, *Jaya: An Advanced Optimization Algorithm and its Engineering Applications* (Springer International Publishing, Cham, 2019). https://doi.org/10.1007/978-3-319-78922-4

25. K. Yu, J.J. Liang, B.Y. Qu, X. Chen, H. Wang, Parameters identification of photovoltaic models using an improved JAYA optimization algorithm. Energy Convers. Manag. **150**, 742–753 (2017)

26. F.J. Pontes, A.P. de Paiva, P.P. Balestrassi, J.R. Ferreira, M.B. da Silva, Optimization of Radial Basis Function neural network employed for prediction of surface roughness in hard turning process using Taguchi's orthogonal arrays. Expert Syst. Appl. **39**(9), 7776–7787 (2012)

27. H. Aouici, M.A. Yallese, B. Fnides, K. Chaoui, T. Mabrouki, Modeling and optimization of hard turning of X38CrMoV5-1 steel with CBN tool: Machining parameters effects on flank wear and surface roughness. J. Mech. Sci. Technol. **25**(11), 2843–2851 (2011)

28. M. Mia, N.R. Dhar, Optimization of surface roughness and cutting temperature in high-pressure coolant-assisted hard turning using Taguchi method. Int. J. Adv. Manuf. Technol. **88**(1–4), 739–753 (2017)

29. M.W. Azizi, S. Belhadi, M.A. Yallese, T. Mabrouki, J.F. Rigal, Surface roughness and cutting forces modeling for optimization of machining condition in finish hard turning of AISI 52100 steel. J. Mech. Sci. Technol. **26**(12), 4105–4114 (2012)

30. I. Asiltürk, H. Akkuş, Determining the effect of cutting parameters on surface roughness in hard turning using the Taguchi method. Measurement **44**(9), 1697–1704 (2011)

31. K. Bouacha, M.A. Yallese, S. Khamel, S. Belhadi, Analysis and optimization of hard turning operation using cubic boron nitride tool. Int. J. Refract. Metal Hard Mater. **45**, 160–178 (2014)

32. H. Aouici, M.A. Yallese, K. Chaoui, T. Mabrouki, J.F. Rigal, Analysis of surface roughness and cutting force components in hard turning with CBN tool: Prediction model and cutting conditions optimization. Measurement **45**(3), 344–353 (2012)

33. D. Manivel, R. Gandhinathan, Optimization of surface roughness and tool wear in hard turning of austempered ductile iron (grade 3) using Taguchi method. Measurement **93**, 108–116 (2016)

34. K. Bouacha, M.A. Yallese, T. Mabrouki, J.F. Rigal, Statistical analysis of surface roughness and cutting forces using response surface methodology in hard turning of AISI 52100 bearing steel with CBN tool. Int. J. Refract. Metal Hard Mater. **28**(3), 349–361 (2010)

35. I. Meddour, M.A. Yallese, R. Khattabi, M. Elbah, L. Boulanouar, Investigation and modeling of cutting forces and surface roughness when hard turning of AISI 52100 steel with mixed ceramic tool: cutting conditions optimization. Int. J. Adv. Manuf. Technol. **77**(5–8), 1387–1399 (2015)

36. Y. Karpat, T. Özel, Multi-objective optimization for turning processes using neural network modeling and dynamic-neighborhood particle swarm optimization. Int. J. Adv. Manuf. Technol. **35**(3–4), 234–247 (2007)

37. W.B. Rashid, S. Goel, J.P. Davim, S.N. Joshi, Parametric design optimization of hard turning of AISI 4340 steel (69 HRC). Int. J. Adv. Manuf. Technol. **82**(1–4), 451–462 (2016)

38. B.M. Gopalsamy, B. Mondal, S. Ghosh, Taguchi method and ANOVA: An approach for process parameters optimization of hard machining while machining hardened steel. J. Sci. Ind. Res. **68**, 686–695 (2009)

39. D.P. Selvaraj, P. Chandramohan, M. Mohanraj, Optimization of surface roughness, cutting force and tool wear of nitrogen alloyed duplex stainless steel in a dry turning process using Taguchi method. Measurement **49**, 205–215 (2014)

40. J.T. Horng, N.M. Liu, K.T. Chiang, Investigating the machinability evaluation of Hadfield steel in the hard turning with Al2O3/TiC mixed ceramic tool based on the response surface methodology. J. Mater. Process. Technol. **208**(1–3), 532–541 (2008)

41. Z. Hessainia, A. Belbah, M.A. Yallese, T. Mabrouki, J.F. Rigal, On the prediction of surface roughness in the hard turning based on cutting parameters and tool vibrations. Measurement **46**(5), 1671–1681 (2013)

42. V.N. Gaitonde, S.R. Karnik, J.P. Davim, Multiperformance optimization in turning of free-machining steel using Taguchi method and utility concept. J. Mater. Eng. Perform. **18**(3), 231–236 (2009)

43. S.K. Shihab, Z.A. Khan, A. Mohammad, A.N. Siddiquee, Optimization of surface integrity in dry hard turning using RSM. Sadhana **39**(5), 1035–1053 (2014)

44. A.P. Paiva, E.J. Paiva, J.R. Ferreira, P.P. Balestrassi, S.C. Costa, A multivariate mean square error optimization of AISI 52100 hardened steel turning. Int. J. Adv. Manuf. Technol. **43**(7–8), 631–643 (2009)

45. S.S. Mahapatra, A. Patnaik, P.K. Patnaik, Parametric analysis and optimization of cutting parameters for turning operations based on Taguchi method, in *Proceedings of the International Conference on Global Manufacturing and Innovation*, vol. 10 (2006), p. 27

46. A.H. Suhail, N. Ismail, S.V. Wong, N.A. Jalil, Optimization of cutting parameters based on surface roughness and assistance of workpiece surface temperature in turning process. Am. J. Eng. Appl. Sci. **3**(1), 102–108 (2010)

47. S. Chinchanikar, S.K. Choudhury, Effect of work material hardness and cutting parameters on performance of coated carbide tool when turning hardened steel: An optimization approach. Measurement **46**(4), 1572–1584 (2013)

48. A.P. Paiva, J.R. Ferreira, P.P. Balestrassi, A multivariate hybrid approach applied to AISI 52100 hardened steel turning optimization. J. Mater. Process. Technol. **189**(1–3), 26–35 (2007)

49. A. Batish, A. Bhattacharya, M. Kaur, M.S. Cheema, Hard turning: Parametric optimization using genetic algorithm for rough/finish machining and study of surface morphology. J. Mech. Sci. Technol. **28**(5), 1629–1640 (2014)

50. S.R. Das, D. Dhupal, A. Kumar, Study of surface roughness and flank wear in hard turning of AISI 4140 steel with coated ceramic inserts. J. Mech. Sci. Technol. **29**(10), 4329–4340 (2015)

51. M. Mia, P.R. Dey, M.S. Hossain, M.T. Arafat, M. Asaduzzaman, M.S. Ullah, S.T. Zobaer, Taguchi S/N based optimization of machining parameters for surface roughness, tool wear and material removal rate in hard turning under MQL cutting condition. Measurement **122**, 380–391 (2018)

52. M. Mia, N.R. Dhar, Prediction and optimization by using SVR, RSM and GA in hard turning of tempered AISI 1060 steel under effective cooling condition. Neural Comput. Appl. 1–22. https://doi.org/10.1007/s00521-017-3192-4

53. Y. Huang, S.Y. Liang, Modeling of cutting forces under hard turning conditions considering tool wear effect. J. Manuf. Sci. Eng. **127**(2), 262–270 (2005)

54. C. Lahiff, S. Gordon, P. Phelan, PCBN tool wear modes and mechanisms in finish hard turning. Rob. Comput. Integr. Manuf. **23**(6), 638–644 (2007)

55. T. Özel, Y. Karpat, Predictive modeling of surface roughness and tool wear in hard turning using regression and neural networks. Int. J. Mach. Tools Manuf. **45**(4–5), 467–479 (2005)

Index

A
Algorithm, 39, 73, 74, 81, 83–85, 93, 103, 104, 106–108, 110, 111, 115, 117–120, 122–125

C
Ceramic, 5, 11–13, 15–17, 28, 29, 32, 33, 35, 40–42, 57
Circularity, 8, 57–60, 62, 63, 67, 68, 70, 76, 79, 80, 89, 92, 123
Cylindricity, 57–60, 63, 65–67, 70, 76, 79, 80, 89, 93, 119, 123, 125

D
Diamond, 13, 16, 17
Difficult-to-machine, 2, 16, 19, 57
Dry cutting, 5, 6, 40

E
Energy, 2, 6, 8, 11, 13, 41, 43, 44, 55

G
Genetic Algorithm (GA), 35, 73, 83, 85, 93, 104, 115, 118

I
Intelligent, 73

M
Mathematical, 30, 36, 38, 39, 60, 67, 74, 82, 103, 116

Metal matrix composite, 10–12

N
Neural network, 73–78, 80–83, 85, 89, 93, 98, 111, 113, 115

P
Productivity, 1, 2, 7, 15, 19, 25, 31, 41, 53, 55, 58

R
Regression, 36, 55, 60, 61, 63, 65–68, 70, 71, 77, 78, 92, 94

S
Steel, 4, 7–10, 12–16, 25, 27, 29, 30, 57, 63
Superalloy, 10, 15
Surface roughness, 5, 8, 29, 39, 40, 43–45, 57–59, 61–64, 70, 75, 76, 79, 80, 89, 92, 113–115, 123

T
Temperature, 7, 9–17, 27, 30, 31, 35, 40, 41, 64, 76, 104, 115, 124
Titanium, 9, 10, 12, 14, 15
Tool wear, 2, 3, 5–8, 11, 16, 26, 27, 29–31, 40–42, 44, 45, 55, 75, 76, 113–115, 124

© The Author(s), under exclusive license to Springer Nature Switzerland AG 2020
M. Patel G. C. et al., *Machining of Hard Materials*,
Manufacturing and Surface Engineering,
https://doi.org/10.1007/978-3-030-40102-3

129

Printed in the United States
By Bookmasters